T0225189

# Electromagnetic Field Theory

Gerd Mrozynski • Matthias Stallein

# Electromagnetic Field Theory

## A Collection of Problems

With 152 Illustrations

 Springer Vieweg

Gerd Mrozynski,
Matthias Stallein,
Universität Paderborn, Germany

ISBN 978-3-8348-1711-2          ISBN 978-3-8348-2178-2 (eBook)
DOI 10.1007/978-3-8348-2178-2

Library of Congress Control Number: 2012943069

The Deutsche Nationalbibliothek lists this publication in the Deutsche Nationalbibliografie;
detailed bibliographic data are available in the Internet at http://dnb.d-nb.de.

Springer Vieweg
© Vieweg+Teubner Verlag | Springer Fachmedien Wiesbaden 2013

*Cover design*: KünkelLopka GmbH, Heidelberg

Printed on acid-free paper

Springer Vieweg is a brand of Springer DE.
Springer DE is part of Springer Science+Business Media.
www.springer-vieweg.de

# Preface

The theory of electromagnetic fields is an integral part of the curriculum of university courses in Electrical Engineering, Information Systems Engineering, and related areas. Often students have difficulties with this subject, because its wonderful theory is hidden behind a mathematical formalism. Quite a few textbooks with emphasis on various aspects of Maxwell's theory are available and it is not the purpose of this book to add another one. Instead it is an attempt to allow for a deeper understanding of static and dynamic fields by the discussion and calculation of typical problems. For a successful learning progress the reader should at first try to solve the problems independently.

Today it is common practice that engineers use software packages to solve Maxwell's equations numerically. Commercial simulation suites offer convenient user interfaces for the modeling and simulation of complex structures. Often it is not necessary to have specific knowledge of the underlying numerical and physical model. Of course, this is the intention of commercial software, but it makes it impossible to check the results and to estimate the inherent error. Especially scientists should always be aware of the validity of their results and whenever possible a comparison with analytic solutions is recommended.

This book covers most of the fundamental analytic approaches for the calculation of static and dynamic electromagnetic fields. In the first chapter Maxwell's theory and the differential equations for the potentials of the fields are briefly summarized. The description is not complete and should rather serve as a formulary. In the following chapters problems of the classical parts of the electromagnetic field theory and their solutions are presented. Wherever it is useful, field patterns of the analytic solutions have been added.

The current edition is a translation of the German book "Elektromagnetische Feldtheorie — Eine Aufgabensammlung" [13]. It includes minor corrections and two additional problems in chapter 6. This English edition should of course address a broader audience, but also support the upcoming bilingual Bachelor and Master Degree courses in Germany.

Paderborn, August 2011

Gerd Mrozynski
Matthias Stallein

# Contents

# Symbols

| | |
|---|---|
| $\vec{A}$ | Vector potential |
| $a$ | Surface, Distance |
| $\vec{B}$ | Magnetic flux density |
| $C$ | Capacitance, Integration path |
| $c_{ij}$ | Capacitance coefficients |
| $\vec{D}$ | Electric flux density |
| $\vec{E}$ | Electric field |
| $\vec{F}$ | Vector potential, Force |
| $\vec{H}$ | Magnetic field |
| $I, i$ | Electric current |
| $\vec{J}$ | Current density |
| $\vec{K}$ | Current sheet |
| $L_{ik}$ | Inductance |
| $\vec{M}$ | Dipole moment |
| $\vec{m}$ | Dipole moment density |
| $P$ | Power |
| $p_v$ | Power loss density |
| $p_{ij}$ | Potential coefficients |
| $Q$ | Charge |
| $\vec{R}, \vec{r}$ | Position vectors |
| $\vec{S}$ | Poynting vector |
| $t, T$ | Time |
| $u$ | Voltage |
| $U, V$ | Potential |
| $v$ | Volume, Velocity |
| $W$ | Energy |
| $w$ | Energy density, Complex variable |
| $Z$ | Wave impedance |
| | |
| $\alpha$ | Skin constant, Angle |
| $\vec{\beta}$ | Phase constant |
| $\vec{\gamma}$ | Propagation constant |
| $\delta$ | Skin depth, Dirac Delta function |
| $\delta_{mn}$ | Kronecker Delta |
| $\varepsilon$ | Permittivity |

| | |
|---|---|
| $\mu$ | Permeability |
| $\kappa$ | Conductivity |
| $\varrho$ | Volume charge density, Coordinate |
| $\sigma$ | Surface charge density, Step function |
| $\lambda_q$ | Line charge density |
| $\lambda$ | Wavelength |
| $\Psi_{e,m}$ | Electric/Magnetic flux |
| $\varphi, \psi$ | Potential function |
| $\Omega$ | Solid angle |
| $\omega$ | Angular frequency |
| $\Phi$ | Radiation pattern, Potential |

$\varepsilon_0 = 8.854 \cdot 10^{-12}$ [As/Vm]

$\mu_0 = 4\pi \cdot 10^{-7}$ [Vs/Am]

Complex quantities are underlined, except of the complex variable $z = x + jy, w = u + jv$ and special functions with complex arguments.

$\underline{A}^*$ conjugate-complex quantity.

$|\vec{\underline{A}}| = A$ Absolute value of a vector or complex quantity.

$\overline{A}$ Time-average value of a quantity.

Im{} Imaginary part of a complex quantity.

Re{} Real part of a complex quantity.

# 1. Fundamental Equations

In a medium of permittivity $\varepsilon$, permeability $\mu$, and conductivity $\kappa$ Maxwell's equations are in integral form

$$\oint_C \vec{H}\,d\vec{s} = \int_a \left(\vec{J} + \frac{\partial \vec{D}}{\partial t}\right) d\vec{a}\,; \quad \oint_C \vec{E}\,d\vec{s} = -\frac{\partial}{\partial t}\int_a \vec{B}\,d\vec{a}$$

$$\vec{B} = \mu\vec{H}\; ; \quad \vec{D} = \varepsilon\vec{E}$$

$$\oint_a \vec{B}\,d\vec{a} = 0\,; \quad \oint_a \vec{D}\,d\vec{a} = \int_v \varrho\,dv\,; \quad \oint_a \vec{J}\,d\vec{a} + \frac{\partial}{\partial t}\int_v \varrho\,dv = 0\,; \quad \vec{J} = \kappa\vec{E}.$$

Here $\vec{H}(\vec{r},t)$ is the magnetic field, $\vec{B}(\vec{r},t)$ is the magnetic flux density, $\vec{E}(\vec{r},t)$ is the electric field, and $\vec{D}(\vec{r},t)$ is the electric flux density.

All fields depend on the space variable $\vec{r}$ and the time $t$ and have to be integrated over surfaces $a$ with a contour $C$ or over surfaces $a$ enclosing the volume $v$. The current density $\vec{J}(\vec{r},t)$ and the charge density $\varrho(\vec{r},t)$ satisfy the continuity equation, which implies the conservation of charge.

In differential form Maxwell's equations are

$$\operatorname{rot}\vec{H} = \vec{J} + \frac{\partial \vec{D}}{\partial t}\,; \quad \operatorname{rot}\vec{E} = -\frac{\partial \vec{B}}{\partial t}\,; \quad \vec{B} = \mu\vec{H}\,; \quad \vec{D} = \varepsilon\vec{E}$$

$$\operatorname{div}\vec{B} = 0\,; \quad \operatorname{div}\vec{D} = \varrho\,; \quad \operatorname{div}\vec{J} + \partial\varrho/\partial t = 0\,; \quad \vec{J} = \kappa\vec{E}.$$

The force $\vec{F}$ of the electromagnetic field on a charge with velocity $\vec{v}$, namely the Lorentz force, is given by

$$d\vec{F} = \varrho(\vec{E} + \vec{v} \times \vec{B})\,dv.$$

For a surface $a$ with normal $\vec{n}$ directed from domain (1) with material properties $\varepsilon_1, \mu_1, \kappa_1$ into domain (2) with material properties $\varepsilon_2, \mu_2, \kappa_2$ the following boundary conditions can be deduced

$$\vec{n} \times \left(\vec{E}_2 - \vec{E}_1\right)\Big|_a = 0\,; \quad \vec{n} \times \left(\vec{H}_2 - \vec{H}_1\right)\Big|_a = 0,\;(\vec{K})$$

$$\vec{n} \left(\vec{B}_2 - \vec{B}_1\right)\Big|_a = 0\,; \quad \vec{n} \left(\vec{D}_2 - \vec{D}_1\right)\Big|_a = \sigma\,; \quad \vec{n} \left(\vec{J}_2 - \vec{J}_1\right)\Big|_a = -\frac{\partial\sigma}{\partial t}\,.$$

Here $\sigma$ is a space- and time-dependent surface charge density within the boundary layer $a$.

If a conducting material is effected by a rapidly time-varying field, then the induced current density decays rapidly from the surface to the inside. The integral of the current density over the coordinate normal to the surface is considered as space-varying current sheet $\vec{K}$, and the boundary condition is

$$\vec{n} \times \vec{H}\Big|_a = \vec{K}.$$

If the surface $a$ carries an impressed current sheet $\vec{K}$, then the relation between the tangential components of the magnetic field in the adjoining subspaces is

$$\vec{n} \times \left(\vec{H}_2 - \vec{H}_1\right)\Big|_a = \vec{K}.$$

The following equations hold in regions with sectional homogeneous material properties, that are characterized by scalar parameters $\varepsilon, \mu$, and $\kappa$.

**Electrostatic Fields**

The equations for the electric field of static charges are

$$\operatorname{rot} \vec{E} = 0; \quad \operatorname{div} \vec{D} = \varrho; \quad \vec{D} = \varepsilon \vec{E}$$

$$\vec{E} = -\operatorname{grad} V; \quad \Delta V = -\varrho/\varepsilon.$$

The electric field at coordinate $\vec{r}_p$ is given by the gradient of the scalar electrostatic potential $V(\vec{r}_p)$. Charge distributions are classified as space charge densities $\varrho(\vec{r}_q)$ in a volume $v$, surface charge densities $\sigma(\vec{r}_q)$ on a surface $a$, or line charge densities $\lambda_q(\vec{r}_q)$ on a line with contour $C$. In a homogeneous space of permittivity $\varepsilon$ the charge distributions possess the following potentials and fields at point $\vec{r}_p$.

$$V(\vec{r}_p) = \frac{1}{4\pi\varepsilon} \int_v \frac{\varrho(\vec{r}_q)}{r} dv; \quad \vec{E}(\vec{r}_p) = -\operatorname{grad} V(\vec{r}_p) = \frac{1}{4\pi\varepsilon} \int_v \varrho(\vec{r}_q) \frac{\vec{r}}{r^3} dv$$

$$V(\vec{r}_p) = \frac{1}{4\pi\varepsilon} \int_a \frac{\sigma(\vec{r}_q)}{r} da; \quad \vec{E}(\vec{r}_p) = -\operatorname{grad} V(\vec{r}_p) = \frac{1}{4\pi\varepsilon} \int_v \sigma(\vec{r}_q) \frac{\vec{r}}{r^3} da$$

$$V(\vec{r}_p) = \frac{1}{4\pi\varepsilon} \int_C \frac{\lambda_q(\vec{r}_q)}{r} ds; \quad \vec{E}(\vec{r}_p) = -\operatorname{grad} V(\vec{r}_p) = \frac{1}{4\pi\varepsilon} \int_C \lambda_q(\vec{r}_q) \frac{\vec{r}}{r^3} ds$$

with $\vec{r} = \vec{r}_p - \vec{r}_q, |\vec{r}| = r$. Similarly the field of $n$ point charges $Q_i$ at positions $\vec{r}_{qi}$; $i = 1, 2, 3, \ldots, n$ is given by

$$V(\vec{r}_p) = \frac{1}{4\pi\varepsilon} \sum_{i=1}^{n} \frac{Q_i}{|\vec{r}_p - \vec{r}_{qi}|}; \quad \vec{E}(\vec{r}_p) = -\operatorname{grad} V(\vec{r}_p) = \frac{1}{4\pi\varepsilon} \sum_{i=1}^{n} Q_i \frac{\vec{r}_p - \vec{r}_{qi}}{|\vec{r}_p - \vec{r}_{qi}|^3}.$$

Line- and point charges are quantities that were introduced to simplify the field computation, but due to the unbounded energy of the fields they are just theoretical models.

The integral of the electric flux density $\vec{D}$ over a surface $a$ leads to the electric Flux

$$\Psi_e = \int_a \vec{D}\, d\vec{a}\,.$$

The integral of $\vec{D}$ over a closed surface $a$, that encloses the volume $v$, is in accordance with the field equation identical to the enclosed charge $Q$.

$$\Psi_e = \oint_a \vec{D}\, d\vec{a} = Q = \int_v \varrho\, dv$$

If the electric flux that runs through a tube is constant $\Psi_e = \text{const}$, then the electric field is parallel to the side surface of the tube. Thus the determination of these flux tubes also defines electric lines of force. In this way often lines of force are obtained easier than by solving the corresponding differential equation

$$d\vec{s} \times \vec{E} = 0\,.$$

Electric dipoles with a moment $\vec{M}(\vec{r}_q)$ in a homogeneous space of permittivity $\varepsilon$ excite the potential $V(\vec{r}_p)$ and the electric field $\vec{E}(\vec{r}_p)$.

$$V(\vec{r}_p) = \frac{\vec{M}(\vec{r}_q)}{4\pi\varepsilon}\,\frac{\vec{r}}{r^3}\,;\quad \vec{E}(\vec{r}_p) = \frac{1}{4\pi\varepsilon}\,\frac{1}{r^3}\left[3\left(\vec{M}\,\vec{r}/r\right)\vec{r}/r - \vec{M}\right];\quad \vec{r} = \vec{r}_p - \vec{r}_q$$

A dipole layer with density $m(\vec{r}_q)$ on the surface $a$ causes the potential

$$V(\vec{r}_p) = \frac{1}{4\pi\varepsilon}\int_a \vec{m}(\vec{r}_q)\vec{r}/r^3\, da\,;\quad \vec{m}(\vec{r}_q) = \vec{n}(\vec{r}_q)\, m(\vec{r}_q)\,.$$

The transition trough a dipole layer in the direction of the surface normal $\vec{n}(\vec{r}_q)$ from region (1) to region (2) induces a change of $m/\varepsilon$ in the potential and for the electric field applies

$$\vec{n} \times \left(\vec{n} \times \left(\vec{E}_2 - \vec{E}_1\right)\right)\Big|_a = 1/\varepsilon\,\text{grad}\left(m(\vec{r}_q)\right)\,;\quad V_2 - V_1\big|_a = m(\vec{r}_q)/\varepsilon\,.$$

If the dipoles have a homogeneous density $\vec{m}_0 = \vec{n}(\vec{r}_q)\, m_0$, then the potential is determined by means of the solid angle $\Omega$, which is given by the border contour of the surface.

$$V(\vec{r}_p) = \frac{m_0}{4\pi\varepsilon}\int_a \vec{n}(\vec{r}_q)\vec{r}/r^3\, da = -\frac{m_0}{4\pi\varepsilon}\Omega\,;\quad \Omega = -\int_a \vec{n}(\vec{r}_q)\,\vec{r}/r^3\, da$$

For spatial dipole distributions $\vec{m}_V(\vec{r}_q)$ in a volume $v$ results in analogy

$$V(\vec{r}_p) = \frac{1}{4\pi\varepsilon} \int_v \vec{m}_V(\vec{r}_q)\,\vec{r}/r^3\,dv.$$

The field of homogeneous distributions $\vec{m}_V = \vec{e}\,m_{V_0}$ can also be calculated by consideration of an equivalent surface charge $\sigma = \vec{n}\,\vec{e}\,m_{V_0}$ on the surface $a$ of the volume $v$, where the total charge on the surface is zero.

Plane electrostatic problems can be treated by a complex potential

$$\underline{P}_e(z) = P_{er}(x,y) + j\,P_{ei}(x,y) = P_{er}(\varrho,\varphi) + j\,P_{ei}(\varrho,\varphi)$$

$$\text{with} \quad \Delta\,P_{er} = 0\,; \quad \Delta\,P_{ei} = 0\,; \quad z = x + jy = \varrho\exp(j\varphi)\,.$$

The Laplace equation is invariant concerning conformal mapping.

For a homogeneous line charge $\lambda_q$ at position $z_q$ the complex potential is

$$\underline{P}_e(z) = -\frac{\lambda_q}{2\pi\varepsilon}\ln\frac{(z-z_q)}{c} = V(\varrho,\varphi) - j/\varepsilon\Psi_e/l =$$

$$= -\frac{\lambda_q}{2\pi\varepsilon}\left(\ln(\varrho/c) + j\arg\frac{(z-z_q)}{c}\right)$$

$$\Rightarrow \quad \underline{E}(z) = -\left(\frac{d\underline{P}_e(z)}{dz}\right)^* = \frac{\lambda_q}{2\pi\varepsilon}\frac{1}{(z-z_q)^*}\,; \quad \varrho = |z - z_q|\,;$$

$$\underline{E}(z) = E_x(x,y) + jE_y(x,y).$$

$\Psi_e/l$ is the electric flux per length $l$ and $c$ is an arbitrary constant. The potential of other planar source distributions is easily deduced from the above equations.

The energy $W_e$ stored in the electric field is the integral of the energy density $w_e$.

$$w_e = \frac{1}{2}\vec{E}\vec{D}\,; \quad W_e = \int_v w_e\,dv = \frac{1}{2}\int_v \vec{E}\vec{D}\,dv$$

In a system of $n$ conducting and charged bodies a simple linear relation between the potentials $V_i$ and the charges $Q_i, i = 1, 2, \ldots, n$ exists.

$$V_k = \sum_{i=1}^{n} p_{ki}Q_i\,; \quad Q_k = \sum_{i=1}^{n} c_{ki}V_i$$

$$Q_k = C_{k\infty}V_k + \sum_{i=1}^{n} C_{ki}(V_k - V_i)$$

Here $p_{ki}$ and $c_{ki}$ are the potential and capacitance coefficients and $C_{ki}$ are mutual capacitances.

**Stationary Current Density Field**

The field equations are

$$\text{rot}\,\vec{E} = 0\,;\quad \text{div}\,\vec{J} = 0\,;\quad \vec{J} = \kappa\vec{E}\,;\quad \vec{E} = -\text{grad}\,V\,;\quad \Delta V = 0\,.$$

As there is an analogy to the electrostatic field in source free regions, most relations of the previous section remain valid.

**Magnetic Field of Stationary Currents**

The field equations are

$$\text{rot}\,\vec{H} = \vec{J}\,;\quad \text{div}\,\vec{B} = 0\,;\quad \vec{B} = \mu\vec{H}$$

$$\vec{B} = \text{rot}\,\vec{A}\,;\quad \text{div}\,\vec{A} = 0\,;\quad \Delta\vec{A} = -\mu\vec{J}\,.$$

With a given current density $\vec{J}(\vec{r}_q)$ in a volume $v$ a solution for the source-free vector potential $\vec{A}(\vec{r}_p)$ is given by

$$\vec{A}(\vec{r}_p) = \frac{\mu}{4\pi}\int_v \frac{\vec{J}(\vec{r}_q)}{|\vec{r}_p - \vec{r}_q|}\,dv\,.$$

Hence the solution for the magnetic field $\vec{H}$ is

$$\vec{H}(\vec{r}_p) = \frac{\vec{B}}{\mu} = \frac{1}{4\pi}\int_v \vec{J}(\vec{r}_q) \times \frac{\vec{r}}{r^3}\,dv\,;\quad \vec{r} = \vec{r}_p - \vec{r}_q\,.$$

The magnetic Flux $\Psi_m$ through a surface $a$ with the contour $C$ is defined by the integrals

$$\Psi_m = \int_a \vec{B}\,d\vec{a} = \int_a \text{rot}\,\vec{A}\,d\vec{a} = \oint_C \vec{A}\,d\vec{s}\,.$$

As the curl of the magnetic field is zero in space with no currents one can also use a scalar potential $V_m$ for the description of the magnetic field. For a line current the solution is

$$\vec{H} = -\text{grad}\,V_m\,;\quad \Delta V_m = 0\,;\quad V_m(\vec{r}_p) = \frac{1}{4\pi\mu}\int_a \vec{m}_m \frac{\vec{r}}{r^3}\,da\,;\quad \vec{m}_m = \mu\vec{n}I\,.$$

A current $I$ on a thin conductor loop with contour $C$ can be replaced by a magnetic dipole density on an arbitrary surface $a$ enclosed by the contour $C$. When passing the surface the potential changes by the value of the current.

The vector potential is determined by

$$\vec{A}(\vec{r}_p) = \frac{1}{4\pi} \int_a \vec{m}_m \times \frac{\vec{r}}{r^3} da ; \quad \vec{r} = \vec{r}_p - \vec{r}_q .$$

Plane magnetic fields can be calculated by a complex magnetic potential $\underline{P}_m(z)$. With $\Delta P_{mr} = 0$ and $\Delta P_{mi} = 0$ it follows

$$\underline{P}_m(z) = \underline{P}_m(x + jy) = P_{mr}(x, y) + j P_{mi}(x, y) = A(x, y) + j\mu V_m(x, y) .$$

Hence the complex magnetic field $\underline{H}(z)$ is

$$\underline{H}(z) = H_x(x, y) + j H_y(x, y) = -j/\mu (d\underline{P}_m(z)/dz)^* .$$

An example for a complex potential is given by a line current $I$ at position $z_q$.

$$\underline{P}_m(z) = -\frac{\mu I}{2\pi} \ln((z - z_q)/c) = A(\varrho, \varphi) + j\mu V_m(\varrho, \varphi)$$

$$A(\varrho) = -\frac{\mu I}{2\pi} \ln(\varrho/c) ; \quad \varrho = |z - z_q| ; \quad \varphi = \arg((z - z_q)/c)$$

$$V_m = -I \frac{\arg((z - z_q)/c)}{2\pi} ; \quad \underline{H}(z) = j \frac{I}{2\pi} \frac{1}{(z - z_q)^*} = j \frac{I}{2\pi\varrho} \exp(j\varphi)$$

Here $c$ is a proper constant.

The energy $W_m$ stored in the magnetic field corresponds to the integral of the magnetic energy density $w_m$.

$$w_m = \frac{1}{2} \vec{H}\vec{B} ; \quad W_m = \frac{1}{2} \int_v \vec{H}\vec{B} dv = \frac{1}{2} \int_{v_L} \vec{A}\vec{J} dv$$

The second integral over the volume of the conductor $v_L$ assumes a regular behavior of the magnetic field and the vector potential at large distances.

In a system of $n$ current-carrying conductors with current densities $\vec{J}_i$ the energy of the field is given by

$$W_m = \frac{1}{2} \sum_{i=1}^{n} \sum_{k=1}^{n} \int_v \vec{H}_i \vec{B}_k dv = \frac{1}{2} \sum_{i=1}^{n} \sum_{k=1}^{n} \int_{v_{Li}} \vec{J}_i \vec{A}_{ik} dv = \frac{1}{2} \sum_{i=1}^{n} \sum_{k=1}^{n} L_{ik} I_i I_k$$

with the field $\vec{H}_i$ of a single conductor $i$ and the vector-potential $\vec{A}_{ik}$ of the current $\vec{J}_k$ evaluated within the conductor $i$. From this it follows the self-inductances and mutual inductances

$$L_{ik} = \int_v \frac{\vec{H}_i}{I_i} \frac{\vec{B}_k}{I_k} dv = \int_{v_{Li}} \frac{\vec{J}_i}{I_i} \frac{\vec{A}_{ik}}{I_k} dv .$$

For two line currents $I_i$ and $I_k$ with the contours $C_i$ and $C_k$ it follows

$$L_{ik} = \frac{1}{I_k} \oint_{C_i} \vec{A}_{ik} d\vec{s}_i = \frac{\Psi_{m_{ik}}}{I_k} ; \qquad i \neq k .$$

## Time-Varying Electromagnetic Fields

The following description again restricts to subspaces of homogeneous permittivity, permeability, and conductivity.

The field equations in case of impressed electric currents with densities $J_E(\vec{r}_q, t)$ are:

$$\operatorname{rot} \vec{H} = \vec{J}_E + \vec{J} + \frac{\partial \vec{D}}{\partial t} ; \quad \operatorname{rot} \vec{E} = -\frac{\partial \vec{B}}{\partial t} ; \quad \operatorname{div} \vec{B} = 0 ; \quad \operatorname{div} \vec{D} = \varrho_E$$

$$\operatorname{div} \vec{J}_E + \frac{\partial \varrho_E}{\partial t} = 0 ; \quad \vec{B} = \mu \vec{H} ; \quad \vec{D} = \varepsilon \vec{E} ; \quad \vec{J} = \kappa \vec{E} .$$

The electrodynamic potentials $\vec{A}(\vec{r}, t)$ and $\varphi(\vec{r}, t)$ satisfy the differential equations

$$\Delta \vec{A} - \kappa \mu \frac{\partial \vec{A}}{\partial t} - \mu \varepsilon \frac{\partial^2 \vec{A}}{\partial t^2} = -\mu \vec{J}_E ; \quad \Delta \varphi - \kappa \mu \frac{\partial \varphi}{\partial t} - \mu \varepsilon \frac{\partial^2 \varphi}{\partial t^2} = -\frac{\varrho_E}{\varepsilon}$$

$$\operatorname{div} \vec{A} + \kappa \mu \varphi + \mu \varepsilon \frac{\partial \varphi}{\partial t} = 0 ; \quad \vec{B} = \operatorname{rot} \vec{A} ; \quad \vec{E} = -\frac{\partial \vec{A}}{\partial t} - \operatorname{grad} \varphi .$$

In case of a sinusoidal time dependency of the field the corresponding equations for the complex amplitudes are

$$\Delta \underline{\vec{A}} - j\omega\kappa\mu\underline{\vec{A}} + \omega^2\mu\varepsilon\underline{\vec{A}} = -\mu\underline{\vec{J}}_E ; \quad \Delta \underline{\varphi} - j\omega\kappa\mu\underline{\varphi} + \omega^2\mu\varepsilon\underline{\varphi} = -\frac{\underline{\varrho}_E}{\varepsilon}$$

$$\operatorname{div} \underline{\vec{A}} + \kappa\mu\underline{\varphi} + j\omega\mu\varepsilon\underline{\varphi} = 0 ; \quad \underline{\vec{B}} = \operatorname{rot} \underline{\vec{A}} ; \quad \underline{\vec{E}} = -j\omega\underline{\vec{A}} - \operatorname{grad} \underline{\varphi} .$$

The Poynting theorem is the law of energy conservation for a volume $v$ surrounded by the surface $a$

$$\underline{P}_s = \overline{P}_{ve} + 2j\omega \left( \overline{W}_m - \overline{W}_e \right) + \oint_a \underline{\vec{S}} d\vec{a} ; \quad \underline{\vec{S}} = \frac{1}{2} \underline{\vec{E}} \times \underline{\vec{H}}^* .$$

Here $\underline{P}_s$ is the complex power of all sources in $v$, $\overline{W}_m$ and $\overline{W}_e$ are the time-averaged energies stored in the electric and magnetic field

$$\overline{W}_m = \frac{1}{4} \operatorname{Re} \left\{ \int_v \underline{\vec{H}}\, \underline{\vec{B}}^* dv \right\} = \frac{\mu}{4} \int_v |\underline{\vec{H}}|^2 dv$$

$$\overline{W}_e = \frac{1}{4}\mathrm{Re}\left\{\int\limits_v \vec{\underline{E}}\,\vec{\underline{D}}^* dv\right\} = \frac{\varepsilon}{4}\int\limits_v |\vec{\underline{E}}|^2 dv$$

and $\overline{P}_{ve}$ is the time-averaged power loss in the volume $v$

$$\overline{P}_{ve} = \frac{1}{2}\mathrm{Re}\left\{\int\limits_v \vec{\underline{E}}\,\vec{\underline{J}}^* dv\right\} = \frac{\kappa}{2}\int\limits_v |\vec{\underline{E}}|^2 dv\,.$$

$\vec{\underline{S}}$ is called the complex Poynting vector.

Within source-free regions also

$$\overline{P}_{ve} = -\frac{1}{2}\mathrm{Re}\left\{\oint\limits_a \vec{\underline{E}} \times \vec{\underline{H}}^* d\vec{a}\right\}$$

holds.

**Quasi Stationary Fields**

The contribution of the displacement current $\partial\vec{D}/\partial t$ to the magnetic field is negligible in case of slowly time-varying fields compared to that of the conduction current. Therefore the field equations are

$$\mathrm{rot}\,\vec{H} = \vec{J}; \quad \mathrm{rot}\,\vec{E} = -\frac{\partial\vec{B}}{\partial t}$$

$$\mathrm{div}\,\vec{B} = 0; \quad \mathrm{div}\,\vec{J} = 0; \quad \vec{B} = \mu\vec{H}; \quad \vec{J} = \kappa\vec{E}$$

$$\vec{B} = \mathrm{rot}\,\vec{A}; \quad \Delta\vec{A} = \kappa\mu\frac{\partial\vec{A}}{\partial t}; \quad \Delta\varphi = \kappa\mu\frac{\partial\varphi}{\partial t}; \quad \mathrm{div}\,\vec{A} + \kappa\mu\varphi = 0$$

$$\vec{E} = -\frac{\partial\vec{A}}{\partial t} - \mathrm{grad}\,\varphi\,.$$

**Electromagnetic Waves**

The fields of <u>plane waves</u> in a homogeneous space of permittivity $\varepsilon$ and permeability $\mu$ depend only on one linear coordinate, here $z$, and on the time $t$ (wave impedance $Z$, phase velocity $v$, wavelength $\lambda$).

$$\vec{E}_{1,2}(\vec{r},t) = \vec{e}E_{1,2}(z \mp vt); \quad \vec{H}_{1,2}(\vec{r},t) = \pm\frac{1}{Z}\vec{e}_z \times \vec{E}_{1,2}(\vec{r},t)$$

$$\vec{e}_z \cdot \vec{E}_{12} = 0; \quad \vec{e}_z\vec{H}_{12} = 0; \quad Z = \sqrt{\frac{\mu}{\varepsilon}}; \quad v = \frac{1}{\sqrt{\mu\varepsilon}}$$

With a sinusoidal time dependency of the angular frequency $\omega$ the solutions are

$$\vec{E}_{1,2}(z,t) = \mathrm{Re}\left\{\vec{\underline{E}}_{1,2}\exp\big(j(\omega t \mp \beta z)\big)\right\}$$

$$\vec{H}_{1,2}(z,t) = \mathrm{Re}\left\{\underline{\vec{H}}_{1,2}\exp(j(\omega t \mp \beta z))\right\}$$

where $\underline{\vec{E}}_{1,2}$ and $\underline{\vec{H}}_{1,2}$ are complex amplitudes with

$$\underline{\vec{H}}_{1,2} = \pm\frac{1}{Z}\vec{e}_z \times \underline{\vec{E}}_{1,2}; \quad \underline{\vec{E}}_{1,2} = \pm Z\,\underline{\vec{H}}_{1,2} \times \vec{e}_z; \quad \beta^2 = \omega^2\mu\varepsilon = (2\pi/\lambda)^2.$$

The Poynting vector has a time-averaged value of

$$\vec{S}_{1,2-} = \pm\vec{e}_z\frac{1}{2}Z|\underline{\vec{H}}_{1,2}|^2 = \pm\vec{e}_z\,\frac{1}{2Z}|\underline{\vec{E}}_{1,2}|^2$$

and a periodical part with the frequency $2\omega$

$$\vec{S}_{1,2\sim} = \pm\frac{1}{2}\,\mathrm{Re}\left\{\underline{\vec{E}}_{1,2} \times \frac{1}{Z}\left(\vec{e}_z \times \underline{\vec{E}}_{1,2}\right)\exp(2j(\omega t \mp \beta z))\right\}$$

$$= \pm\vec{e}_z\frac{1}{2Z}\,\mathrm{Re}\left\{\underline{\vec{E}}_{1,2}^2\exp(2j(\omega t \mp \beta z))\right\}$$

$$= \pm\vec{e}_z\frac{1}{2}Z\,\mathrm{Re}\left\{\underline{\vec{H}}_{1,2}^2\exp(2j(\omega t \mp \beta z))\right\}.$$

In <u>source-free spaces</u> the fields are derivable from two potentials, which describe transverse electric (TE-) or transverse magnetic (TM-)fields with respect to the direction vector $\vec{e}$. For the complex amplitudes applies

$$\underline{\vec{A}}_{TM} = \vec{e}\underline{A}_{TM}; \quad \underline{\vec{F}}_{TE} = \vec{e}\underline{F}_{TE}$$

$$\Delta\underline{F}_{TE} + \beta^2\underline{F}_{TE} = 0; \quad \Delta\underline{A}_{TM} + \beta^2\underline{A}_{TM} = 0$$

$$\underline{\vec{H}} = \underline{\vec{H}}_{TE} + \underline{\vec{H}}_{TM} = \frac{1}{\mu}\mathrm{rot}\,\underline{\vec{A}}_{TM} + \frac{1}{j\omega\mu\varepsilon}\mathrm{rot}\,\mathrm{rot}\,\underline{\vec{F}}_{TE}$$

$$\underline{\vec{E}} = \underline{\vec{E}}_{TE} + \underline{\vec{E}}_{TM} = -\frac{1}{\varepsilon}\mathrm{rot}\,\underline{\vec{F}}_{TE} + \frac{1}{j\omega\mu\varepsilon}\mathrm{rot}\,\mathrm{rot}\,\underline{\vec{A}}_{TM}; \quad \beta^2 = \omega^2\mu\varepsilon; \quad \kappa = 0.$$

A <u>Hertzian dipole</u> located at the point of origin with the moment $\underline{\vec{M}} = \vec{e}_z\underline{M} = \vec{e}_z\underline{I}s$ and $s \ll \lambda$ possesses in the far-field region $r \gg \lambda$ the field

$$\underline{\vec{H}}(\vec{r}) = \frac{j\beta}{4\pi}\left(\underline{\vec{M}} \times \frac{\vec{r}}{r}\right)\frac{\exp(-j\beta r)}{r}; \quad \beta = \omega\sqrt{\mu\varepsilon} = \frac{2\pi}{\lambda}$$

$$\underline{\vec{E}}(\vec{r}) = \frac{\beta^2}{j\omega 4\pi\varepsilon}\frac{\vec{r}}{r} \times \left(\underline{\vec{M}} \times \frac{\vec{r}}{r}\right)\frac{\exp(-j\beta r)}{r} = Z\,\underline{\vec{H}} \times \frac{\vec{r}}{r}.$$

The radiation pattern $\Phi$ is given by

$$\Phi = \frac{Z}{8}\left(\frac{|\underline{\vec{M}}|}{\lambda}\right)^2\sin^2\vartheta; \quad \cos\vartheta = \vec{e}_z\frac{\vec{r}}{r}.$$

A <u>linear antenna</u> with the given current distribution $\underline{I}(\vec{r}_q)$ on a line with contour $C$ leads to the following field in the far-field region:

$$\underline{\vec{H}}(\vec{r}_p) = \frac{j\beta}{4\pi} \frac{\exp(-j\beta r_p)}{r_p} \int\limits_C \underline{I}(\vec{r}_q) \exp\left(j\beta\vec{r}_q \frac{\vec{r}_p}{r_p}\right) \left(d\vec{s}_q \times \frac{\vec{r}_p}{r_p}\right)$$

$$\underline{\vec{E}}(\vec{r}_p) = Z\underline{\vec{H}} \times \frac{\vec{r}_p}{r_p}; \quad Z = \sqrt{\frac{\mu}{\varepsilon}}; \quad \vec{r} = \vec{r}_p - \vec{r}_q; \quad r_p = |\vec{r}_p|.$$

<u>TEM-waves</u> guided by an ideal <u>transmission line</u> in $z$-direction are derived by means of the potentials

$$\underline{\vec{A}} = \vec{e}_z \Phi(x,y) \exp(\mp j\beta z); \quad \underline{\vec{E}} = \mp\frac{1}{\sqrt{\mu\varepsilon}} \operatorname{grad} \Phi(x,y) \exp(\mp j\beta z)$$

$$\underline{\vec{H}} = \frac{1}{\mu} \operatorname{rot} \underline{\vec{A}} = \frac{1}{\mu} \operatorname{grad} \Phi(x,y) \times \vec{e}_z \exp(\mp j\beta z) = \pm\frac{1}{Z}\vec{e}_z \times \underline{\vec{E}}$$

$$\Delta\Phi(x,y) = 0; \quad \beta = \omega\sqrt{\mu\varepsilon}.$$

The integrals of the fields bring out the voltage $u(z,t)$ and the current $i(z,t)$ of the transmission line

$$\frac{\partial^2 u}{\partial z^2} = L'C'\frac{\partial^2 u}{\partial t^2}; \quad \frac{\partial^2 i}{\partial z^2} = L'C'\frac{\partial^2 i}{\partial t^2}; \quad -\frac{\partial u}{\partial z} = L'\frac{\partial i}{\partial t}; \quad -\frac{\partial i}{\partial z} = C'\frac{\partial u}{\partial t}$$

where $L'$ is the inductance per unit length and $C'$ is the capacitance per unit length in $z$-direction. The solutions are

$$u(z,t) = u_{1,2}(z \mp vt); \quad i(z,t) = i_{1,2}(z \mp vt); \quad v = \frac{1}{\sqrt{L'C'}}$$

$$Z = \sqrt{\frac{L'}{C'}}; \quad u_1 = Z\,i_1; \quad u_2 = -Z\,i_2.$$

<u>TE- and TM-waves</u> within ideal <u>cylindrical waveguides</u> in $z$-direction are described by the potentials $\underline{\vec{A}}_{TM}$ and $\underline{\vec{A}}_{TE}$.

$$\underline{\vec{A}}_{TM} = \vec{e}_z \underline{A}_{TM} = \vec{e}_z \underline{U}_{TM}(u_1, u_2) \exp\left(\mp j\beta_z^{TM} z\right)$$

$$\underline{\vec{F}}_{TE} = \vec{e}_z \underline{F}_{TE} = \vec{e}_z \underline{U}_{TE}(u_1, u_2) \exp\left(\mp j\beta_z^{TE} z\right)$$

$$\Delta\underline{U}_{TE,TM} + \left(\beta^2 - (\beta_z^{TE,TM})^2\right) \underline{U}_{TE,TM} = 0$$

Here $(u_1, u_2)$ are the orthogonal curvilinear coordinates in the plane $z = \text{const}$.

At the perfect conducting boundaries of the waveguide with the contour $C$ the boundary conditions

$$\underline{U}_{TM}|_C = 0 \; ; \qquad \left.\frac{\partial \underline{U}_{TE}}{\partial n}\right|_C = 0$$

have to be satisfied. Thereby the phase constants $\beta_z^{TE}$ and $\beta_z^{TM}$ can be identified.

If the <u>source distributions</u> $\vec{J}_E(\vec{r}_q, t)$ and $\varrho_E(\vec{r}_q, t)$ are given in a homogeneous space, then the fields can be deduced from retarded potentials.

$$\vec{A}(\vec{r}_p, t) = \frac{\mu}{4\pi} \int\limits_v \frac{\vec{J}_E(\vec{r}_q, t^*)}{|\vec{r}_p - \vec{r}_q|} \, dv$$

$$\varphi(\vec{r}_p, t) = \frac{1}{4\pi\varepsilon} \int\limits_v \frac{\varrho_E(\vec{r}_q, t^*)}{|\vec{r}_p - \vec{r}_q|} \, dv$$

Therein $t^* = t - r/v_{ph}$ is the retarded time with the phase velocity $v_{ph} = 1/\sqrt{\mu\varepsilon}$.

In case of a sinusoidal time dependency the complex amplitudes are given by

$$\underline{\vec{A}}(\vec{r}_p) = \frac{\mu}{4\pi} \int\limits_v \underline{\vec{J}}_E(\vec{r}_q) \frac{\exp(-j\beta|\vec{r}_p - \vec{r}_q|)}{|\vec{r}_p - \vec{r}_q|} \, dv$$

$$\underline{\varphi}(\vec{r}_p) = \frac{1}{4\pi\varepsilon} \int\limits_v \underline{\varrho}_E(\vec{r}_q) \frac{\exp(-j\beta|\vec{r}_p - \vec{r}_q|)}{|\vec{r}_p - \vec{r}_q|} \, dv$$

with $\beta = \omega\sqrt{\mu\varepsilon}$.

# 2. Electrostatic Fields

## 2.1 Charged Concentric Spheres

Consider two conducting hollow spheres of radius $r = b$ and $r = c > b$. The inner sphere contains a spherical space charge with density $\varrho$ [As/m$^3$] and radius $a < b - e$, which is eccentrically positioned at a distance $e$ from the center. The homogeneous permittivity in $b < r < c$ is $\varepsilon$, while the permittivity in the remaining space is $\varepsilon_0$. The potential on the outer sphere at $r = c$ is $V_0$.

Find the potential on the sphere $r = b$, when it is uncharged.

Now the charge on the sphere $r = b$ is set to $Q_1$. What is the Potential on the surface and what is the charge $Q_0$ of the sphere at $r = c$?

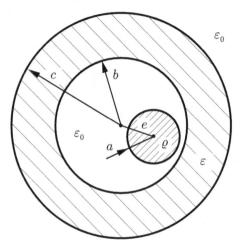

In $b < r < c$ the electric field is of spherical symmetry and is described by the potential $V(r)$.

$$V(r) = C_1 + C_2/r ; \quad \vec{E} = -\operatorname{grad} V = -\vec{e}_r \frac{\partial V}{\partial r} = \vec{e}_r E(r)$$

$$E(r) = C_2/r^2 ; \quad D(r) = \varepsilon E(r) = \varepsilon C_2/r^2$$

If the sphere $r = b$ is uncharged, the electrical flux through a surface $b < r = \text{const} < c$ is equal to the charge quantity in the region $r < b$.

$$D(r)\, 4\pi r^2 = \varrho\, 4/3\, \pi a^3 = Q = 4\pi\varepsilon\, C_2 \quad \Rightarrow \quad C_2 = \frac{Q}{4\pi\varepsilon}$$

Hence the potential is

$$V(r) = C_1 + \frac{Q}{4\pi\varepsilon}\frac{1}{r} ; \quad b \le r \le c ; \qquad V(r) = V_0 \frac{c}{r} ; \quad r \ge c .$$

Furthermore the potential at $r = c$ is $V_0$

$$V(r = c) = V_0 = C_1 + \frac{Q}{4\pi\varepsilon}\frac{1}{c} \quad \Rightarrow \quad C_1 = V_0 - \frac{Q}{4\pi\varepsilon}\frac{1}{c},$$

thus

$$V(r) = V_0 + \frac{Q}{4\pi\varepsilon}\left(\frac{1}{r} - \frac{1}{c}\right); \qquad b \le r \le c.$$

With a vanishing total charge the sphere $r = b$ is on the potential

$$V(r = b) = V_0 + \frac{Q}{4\pi\varepsilon}\left(\frac{1}{b} - \frac{1}{c}\right).$$

If the total charge of the sphere $r = b$ is $Q_1$, then the potential is

$$V(r) = V_0 + \frac{Q + Q_1}{4\pi\varepsilon}\left(\frac{1}{r} - \frac{1}{c}\right); \qquad b \le r \le c; \qquad V(r) = V_0\frac{c}{r}; \qquad r \ge c.$$

The sphere $r = c$ bears on its inner surface the total charge of $-(Q+Q_1)$ and the outer side carries a homogeneous surface charge density $\sigma$:

$$\sigma = D(r)|_{\substack{r \ge c \\ r \to c}} = -\varepsilon_0 \left.\frac{\partial V}{\partial r}\right|_{\substack{r \ge c \\ r \to c}} = \varepsilon_0 \frac{V_0}{c}.$$

Therefore the total charge of the sphere $r = c$ is

$$Q_0 = -(Q + Q_1) + 4\pi\,\varepsilon_0\,cV_0.$$

Finally, it should be noted that the position of the charge density $\varrho$ has no influence on the potential in $\varrho > b$.


## 2.2 Mutual Capacitances of a Screened Parallel-Wire Line

A parallel-wire line consists of ideal conductors with a distance $2b$ and a radius $d \ll b$. It is symmetrically surrounded by a perfectly conducting circular screen of radius $a > b$ with $a - b \gg d$. The permittivity inside the screen is $\varepsilon$.

Calculate the potential and the mutual capacitances per longitudinal unit length inside the screen.

The electric field in the complex plane $z = x + jy = \varrho\exp(j\varphi)$ is equivalently approximated by the field of four line charges. The line charges $\lambda_{q1}$ at $z = b$ and $\lambda_{q2}$ at $z = -b$ have to be mirrored at the cylinder $\varrho = a$. Hence there are additional charges $-\lambda_{q1}$ at $z = a^2/b$ and $-\lambda_{q2}$ at $z = -a^2/b$, and the complex potential $\underline{P}_e(z)$ is

$$\underline{P}_e(z) = -\frac{\lambda_{q1}}{2\pi\varepsilon}\left(\ln\frac{z - b}{a - b} - \ln\frac{z - a^2/b}{a - a^2/b}\right) - \frac{\lambda_{q2}}{2\pi\varepsilon}\left(\ln\frac{z + b}{a + b} - \ln\frac{z + a^2/b}{a + a^2/b}\right),$$

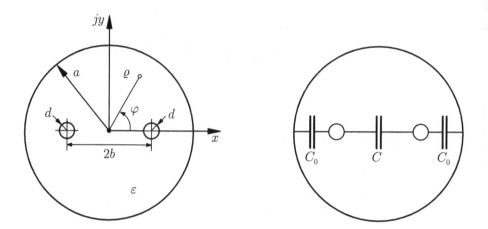

where the real part vanishes on the cylinder $\varrho = a$. Rearranging the arguments of the logarithm functions results in

$$\underline{P}_e(z) = \frac{\lambda_{q1}}{2\pi\varepsilon}\left(\ln\frac{z - a^2/b}{z - b} + \ln\left(\frac{b}{a}\frac{a - b}{b - a}\right)\right) + \frac{\lambda_{q2}}{2\pi\varepsilon}\left(\ln\frac{z + a^2/b}{z + b} + \ln(b/a)\right).$$

For the potential $V(\varrho, \varphi) = \text{Re}\{\underline{P}_e(z)\}$ it follows

$$V(\varrho, \varphi) = \frac{\lambda_{q1}}{2\pi\varepsilon}\ln\left|\frac{z - a^2/b}{z - b}\frac{b}{a}\right| + \frac{\lambda_{q2}}{2\pi\varepsilon}\ln\left|\frac{z + a^2/b}{z + b}\frac{b}{a}\right|.$$

If the potential on the screen is zero, then the potential coefficients $p_{ik}$ for the present problem are deduced from

$$V_1 = p_{11}\lambda_{q1} + p_{12}\lambda_{q2} = V(\varrho = b - d, \varphi = \pi)$$

$$V_2 = p_{21}\lambda_{q1} + p_{22}\lambda_{q2} = V(\varrho = b - d, \varphi = 0).$$

Here it has to be noted, that all potential coefficients refer to the unit length $l$. Now, if we compare these equations with the solution for the potential

$$V_1 = \frac{\lambda_{q1}}{2\pi\varepsilon}\ln\left|\frac{b - a^2/b}{d}\frac{b}{a}\right| + \frac{\lambda_{q2}}{2\pi\varepsilon}\ln\left|\frac{b + a^2/b}{2a}\right|; \quad d \ll b$$

$$V_2 = \frac{\lambda_{q1}}{2\pi\varepsilon}\ln\left|\frac{-b - a^2/b}{-2b}\frac{b}{a}\right| + \frac{\lambda_{q2}}{2\pi\varepsilon}\ln\left|\frac{-b + a^2/b}{d}\frac{b}{a}\right|.$$

we derive

$$p_{11} = p_{22} = \frac{1}{2\pi\varepsilon}\ln\left(b/d(a/b - b/a)\right) = p$$

$$p_{12} = p_{21} = \frac{1}{2\pi\varepsilon}\ln\left((b/a + a/b)/2\right) = p_0$$

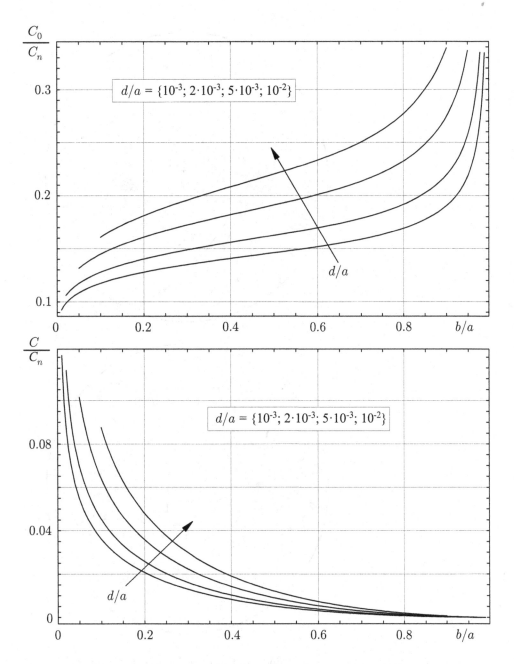

Fig. 2.2–1: Normalized mutual capacitances in dependence on $b/a$ (Distance between conductors $2b$, radius of screen $a$, $C_n = 2\pi\varepsilon$)

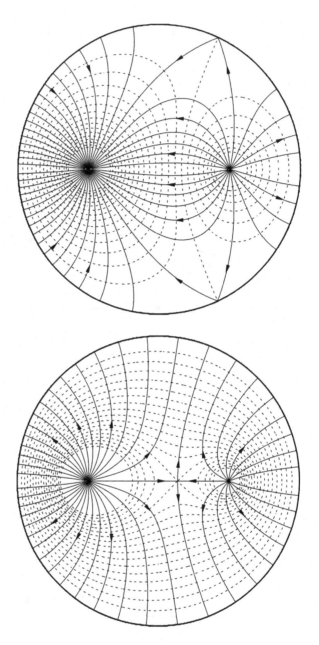

Fig. 2.2–2: Equipotential lines and electric lines of force for $\lambda_{q1}/\lambda_{q2} = -2$ (top) and $\lambda_{q1}/\lambda_{q2} = 2$ (bottom)

and for the potentials we write

$$V_1 = p_0\,\lambda_{q1} + p\,\lambda_{q2}\;;\qquad V_2 = p\,\lambda_{q1} + p_0\,\lambda_{q2}\,.$$

With the capacitance coefficients $c_{ik}$ the equations for the charges read as

$$\lambda_{q1} = c_{11}\,V_1 + c_{12}\,V_2 = \frac{V_1 p_0 - V_2 p}{p_0^2 - p^2}$$

$$\lambda_{q2} = c_{21}\,V_1 + c_{22}\,V_2 = \frac{V_2 p_0 - V_1 p}{p_0^2 - p^2}\,,$$

thus the capacitance coefficients are

$$c_{11} = c_{22} = \frac{p_0}{p_0^2 - p^2} = c_0\;;\qquad c_{12} = c_{21} = \frac{-p}{p_0^2 - p^2} = c.$$

Finally rewriting the equations of the the charges in terms of potential differences

$$\begin{aligned}\lambda_{q1} &= C_0 V_1 + C_{12}(V_1 - V_2) = (c_{11} + c_{12})V_1 - c_{12}(V_1 - V_2)\\ \lambda_{q2} &= C_0 V_2 + C_{21}(V_2 - V_1) = (c_{21} + c_{22})V_2 - c_{21}(V_2 - V_1)\end{aligned}$$

leads to the mutual capacities per unit length

$$C_0 = \frac{1}{p_0 + p}\;;\qquad C_{12} = C_{21} = \frac{p}{p_0^2 - p^2} = C$$

$$C_0 = \frac{2\pi\varepsilon}{\ln\left[b/(2d)\left((a/b)^2 - (b/a)^2\right)\right]}$$

$$C = \frac{2\pi\varepsilon\,\ln\left[1/2\,(a/b + b/a)\right]}{\left[\ln\left((a/b - b/a)\,b/d\right)\right]^2 - \left[\ln\left(1/2\,(a/b + b/a)\right)\right]^2}\,.$$

## 2.3  Singular Points and Lines in the Field of Point Charges

On a straight line in the space of permittivity $\varepsilon$ the point charges $Q_1$ and $Q_2$ are placed in front of the point charge $Q$ with the distances $\vec{r}_1$ and $\vec{r}_2$.

What are the values of the charges $Q_1$ and $Q_2$, in order that a singular point (a singular line) occurs at arbitrary distance $\vec{r}_s$ from $Q$?

The condition for the electric field is:

$$\frac{Q}{4\pi\varepsilon}\frac{\vec{r}_s}{r_s^3} + \frac{Q_1}{4\pi\varepsilon}\frac{\vec{r}_s - \vec{r}_1}{|\vec{r}_s - \vec{r}_1|^3} + \frac{Q_2}{4\pi\varepsilon}\frac{\vec{r}_s - \vec{r}_2}{|\vec{r}_s - \vec{r}_2|^3} = 0\;;\qquad |\vec{r}_s| = r_s$$

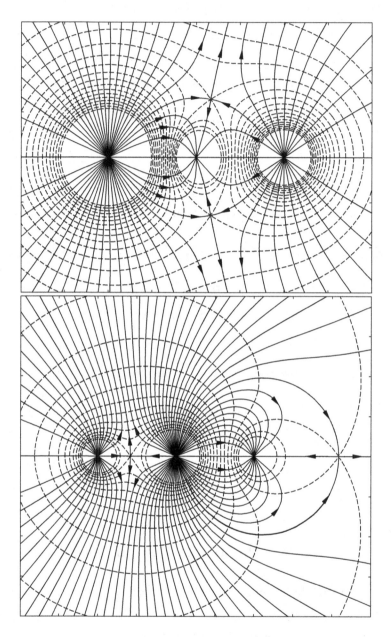

Fig. 2.3–1: Equipotential lines and electric lines of force of the three point charges,
$|\vec{r}_2| = 2|\vec{r}_1|$
top:   $Q_1/Q = -1/4$;   $Q_2/Q = 1/2$
bottom:   $Q_1/Q = 2$;   $Q_2/Q = -2/3$

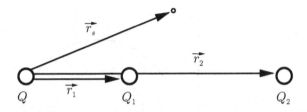

Building the vector and scalar product with $\vec{r}_s$ brings out the equations

$$Q_1 \frac{\vec{r}_s \times \vec{r}_1}{|\vec{r}_s - \vec{r}_1|^3} + Q_2 \frac{\vec{r}_s \times \vec{r}_2}{|\vec{r}_s - \vec{r}_2|^3} = 0$$

and

$$\frac{Q}{r_s} + Q_1 \frac{r_s^2 - \vec{r}_1 \vec{r}_s}{|\vec{r}_s - \vec{r}_1|^3} + Q_2 \frac{r_s^2 - \vec{r}_2 \vec{r}_s}{|\vec{r}_s - \vec{r}_2|^3} = 0.$$

Hence we get the solution:

$$Q_1 = -Q_2 \frac{r_2}{r_1} \left| \frac{\vec{r}_s - \vec{r}_1}{\vec{r}_s - \vec{r}_2} \right|^3 = Q \frac{|\vec{r}_s - \vec{r}_1|^3}{r_s^3} \frac{r_2}{r_1 - r_2}; \qquad Q_2 = Q \frac{|\vec{r}_s - \vec{r}_2|^3}{r_s^3} \frac{r_1}{r_2 - r_1}.$$

## 2.4 Force on a Point Charge by the Field of a Space Charge

A hollow spherical charge with an inner and outer radius of $a$ and $b$ is placed in front of the conducting plane $z = 0$ with its center at $z = c > b$. The charge density is $\varrho$ [As/m³]. On the axis of rotation, normal to the conducting plane and through the center of the sphere, a point charge $Q$ is positioned at $z = d$ with $c - b < d < c - a$. The permittivity $\varepsilon$ is constant.

What is the force on the point charge $Q$?

If there is no conducting plane, then the force $\vec{F}_1 = \vec{e}_z F_1$ on the charge $Q$ is given by

$$\begin{aligned} F_1 = Q\,E_1 &= -Q \varrho \frac{4}{3} \pi \left[ (c-d)^3 - a^3 \right] \frac{1}{4\pi\varepsilon} \frac{1}{(c-d)^2} \\ &= -\frac{Q\varrho}{3\varepsilon} (c-d) \left[ 1 - \left( \frac{a}{c-d} \right)^3 \right]. \end{aligned}$$

Thus the force on the charge $Q$ caused by the hollow spherical charge density is equivalent to that of the total charge within the radii $c-d$ and $a$ located in the center $z = c$. The remaining charge between the radii $b$ and $c-d$ does not exert a force on the point charge.

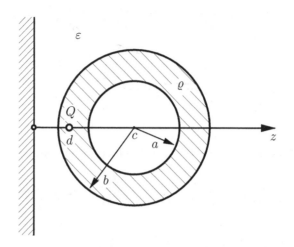

In the presence of the conducting plane additional image charges $-Q$ at $z = -d$ and $-4/3\pi(b^3 - a^3)$ at $z = -c$ have to be regarded. This leads to the additional force $\vec{F}_2 = \vec{e}_z F_2$.

$$
\begin{aligned}
F_2 &= -Q\left[\frac{Q}{4\pi\varepsilon}\frac{1}{4d^2} + \frac{4}{3}\pi\frac{(b^3 - a^3)}{4\pi\varepsilon}\varrho\frac{1}{(d + c)^2}\right] \\
&= -\frac{Q^2}{16\pi\varepsilon}\frac{1}{d^2} - \frac{Q\varrho}{3\varepsilon}\frac{b^3 - a^3}{(d + c)^2}
\end{aligned}
$$

Hence the total force on the charge $Q$ is

$$
\vec{F} = \vec{e}_z\left(F_1 + F_2\right).
$$

## 2.5   Charge Density on a Conducting Cylinder in Front of a Conducting Plane

A conducting cylinder of infinite length and with radius $a$ is located in front of a conducting plane at $x = 0$ with potential $V = 0$. The distance of the cylinder axis parallel to the plane is $b > a$.

Calculate the maximum of the charge density on the cylinder, if it is on the potential $V_0$. The permittivity is $\varepsilon$.

For $x > 0$ outside the cylinder the field is equivalently described by two line charges $\pm\lambda_q$ of infinite extension at positions $(x = \pm c; y = 0)$.

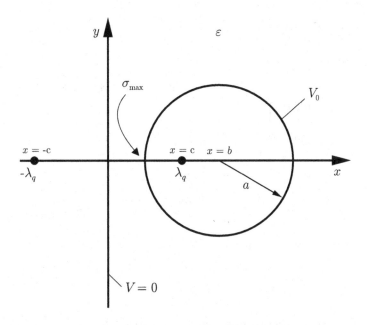

The potential at points $(x \geq 0, y)$ outside the cylinder is

$$V(x,y) = -\frac{\lambda_q}{2\pi\varepsilon} \ln \frac{\sqrt{(x-c)^2 + y^2}}{\sqrt{(x+c)^2 + y^2}}.$$

Now the quantities $\lambda_q$ and $c$ are determined by evaluating the potential $V_0$ on the boundary of the cylinder, e.g. at the points $(x = b - a; y = 0)$ and $(x = b + a; y = 0)$.

$$\ln \left| \frac{b + a - c}{b + a + c} \right| = \ln \left| \frac{b - a - c}{b - a + c} \right|$$

$$\Rightarrow \quad [b + (a - c)] \, [b - (a - c)] = [(a + c) + b] \, [(a + c) - b]$$

$$\Rightarrow \quad b^2 - (a - c)^2 = (a + c)^2 - b^2 \; ; \qquad c^2 = b^2 - a^2$$

Thus the charge density $\lambda_q$ is given by

$$\lambda_q = -\frac{2\pi\varepsilon V_0}{\ln \dfrac{b + a - \sqrt{b^2 - a^2}}{b + a + \sqrt{b^2 - a^2}}} .$$

For the maximum value of $\sigma$ at the point $(x = b - a; y = 0)$ it follows

$$\sigma = -\varepsilon \frac{\partial V}{\partial n} \bigg|_{x=b-a; y=0} = \varepsilon \frac{\partial V}{\partial x} \bigg|_{x=b-a; y=0}$$

$$V(x, y = 0) = -\frac{\lambda_q}{2\pi\varepsilon} \ln \left| \frac{x - c}{x + c} \right| = -\frac{\lambda_q}{2\pi\varepsilon} \ln \frac{c - x}{c + x} ; \quad 0 \leq x \leq b - a$$

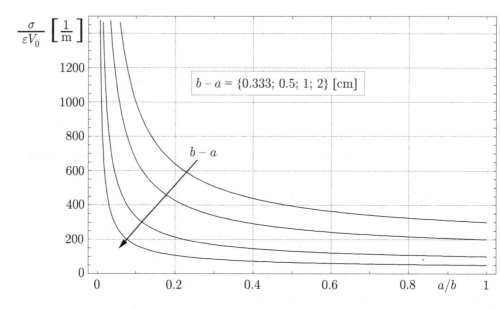

Fig. 2.5–1: Maximum value of $\sigma$ in dependence on the cylinder radius $a$ (the distance $b - a$ is constant)

$$\frac{\partial V}{\partial x}\bigg|_{x=b-a;y=0} = \frac{\lambda_q}{2\pi\varepsilon}\left[\frac{1}{c-x} + \frac{1}{c+x}\right]_{x=b-a} = \frac{\lambda_q}{2\pi\varepsilon}\frac{2c}{c^2 - (b-a)^2}$$

$$\sigma = \frac{\lambda_q}{\pi}\frac{\sqrt{b^2 - a^2}}{(b^2 - a^2) - (b-a)^2} = \frac{\lambda_q}{\pi}\frac{\sqrt{b^2 - a^2}}{2a(b-a)} = \frac{\lambda_q}{2\pi a}\sqrt{\frac{1 + a/b}{1 - a/b}}\,.$$

## 2.6   Potential of Concentric Spheres

Two concentric and conducting spherical surfaces at $r = a$ and $r = b$ are on the potentials $V(r = a) = V_1$ and $V(r = b) = V_2$. The space $a < r < b$ between the spheres has the permittivity $\varepsilon_1$ and the outer space $r > b$ has the permittivity $\varepsilon_2$.

What is the total charge of the conducting surfaces? On what terms does the total charge of the surface $r = b$ vanish?

In $r \leq a$ the potential is constant and the electric field vanishes. Outside of the surface $r = b$ the potential is given in terms of the reciprocal distance $1/r$ (compare with the potential of a point charge or an homogeneous spherical charge). For this reasons the

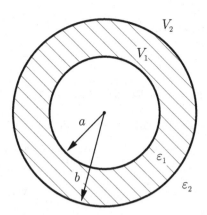

ansatz for the potential is

$$
V(r) = \begin{cases}
V_2\, b/r & ; \quad r \geq b \\
C_1\, a/r + C_2 & ; \quad a \leq r \leq b \\
V_1 & ; \quad r \leq a
\end{cases}
$$

For the determination of the constants $C_1$ and $C_2$ the boundary conditions in $r = a$ and $r = b$ have to be evaluated.

$$
V(r = a) = V_1 = C_1 + C_2 \; ; \qquad V(r = b) = V_2 = C_1\, a/b + C_2
$$

$$
C_1 = \frac{V_1 - V_2}{1 - a/b} \; ; \qquad C_2 = V_1 - \frac{V_1 - V_2}{1 - a/b}
$$

Thus the potential in $a \leq r \leq b$ is given by

$$
V(r) = V_1 - (V_1 - V_2)\, \frac{1 - a/r}{1 - a/b} \, .
$$

In the following $\sigma_1$ is the charge density on the sphere $r = a$. $\sigma_{2i}$ and $\sigma_{2a}$ are the inner $(r < b)$ and outer $(r > b)$ charge densities on the conducting sphere $r = b$.

$$
\sigma_1 = -\varepsilon_1 \left. \frac{\partial V}{\partial r} \right|_{a \leq r \leq b,\, r \to a}
$$

$$
\sigma_{2i} = \varepsilon_1 \left. \frac{\partial V}{\partial r} \right|_{a \leq r \leq b,\, r \to b} \; ; \qquad \sigma_{2a} = -\varepsilon_2 \left. \frac{\partial V}{\partial r} \right|_{r > b,\, r \to b}
$$

$$
\sigma_1 = \varepsilon_1 \frac{C_1}{a} \; ; \qquad \sigma_{2i} = -\varepsilon_1 C_1 \frac{a}{b^2} \; ; \qquad \sigma_{2a} = \varepsilon_2 \frac{V_2}{b}
$$

The total charges are

$$Q_1 = 4\pi a^2 \sigma_1 = 4\pi \varepsilon_1 \frac{a}{1 - a/b}(V_1 - V_2)$$

$$Q_2 = 4\pi b^2(\sigma_{2i} + \sigma_{2a}) = 4\pi \varepsilon_2 b \left[V_2 - \frac{\varepsilon_1/\varepsilon_2 \, a/b}{1 - a/b}(V_1 - V_2)\right].$$

The total charge $Q_2$ vanishes on condition that

$$V_2 = V_1 \frac{\varepsilon_1/\varepsilon_2 \, a/b}{1 - a/b(1 - \varepsilon_1/\varepsilon_2)}$$

$$= V_1 \, a/b \quad \text{with} \quad \varepsilon_1 = \varepsilon_2 \, .$$

In this case the potential $V_2$ relates to the value at points $r = b$, when only the sphere $r = a$ with potential $V_1$ is specified ($V(r) = V_1 \, a/r$ for $r \geq a$).

## 2.7   Dipole within a Dielectric Sphere

An electric dipole of moment $\vec{M} = \vec{e}_z M$ is positioned in the center of a dielectric sphere of permittivity $\varepsilon$ and radius $a$.

Calculate the potential and the field in the sphere $r < a$ and in the outer space $r > a$ of permittivity $\varepsilon_0$. Which surface charge has to be brought onto the sphere $r = a$, in order that the field outside the sphere vanishes?

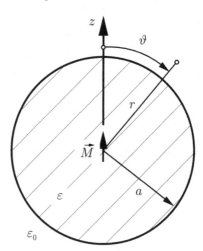

The dipole in homogeneous space of permittivity $\varepsilon$ excites the potential

$$V_e(\vec{r}) = \frac{\vec{M}}{4\pi\varepsilon} \frac{\vec{r}}{r^3} = \frac{M}{4\pi\varepsilon} \frac{\cos\vartheta}{r^2} = V_e(r, \vartheta); \quad \vec{r} = \vec{e}_r r; \quad \vec{M} = \vec{e}_z M \, .$$

If the space $r > a$ is of permittivity $\varepsilon_0$, then a solution of Laplace's equation

$$\Delta V(r, \vartheta) = \frac{\partial^2 V}{\partial r^2} + \frac{2}{r} \frac{\partial V}{\partial r} + \frac{1}{r^2} \frac{1}{\sin \vartheta} \frac{\partial}{\partial \vartheta} \left( \sin \vartheta \frac{\partial V}{\partial \vartheta} \right) = 0$$

has to be added to the potential $V_e$ in both spaces $r < a$ and $r > a$ respectively. With the substitution $u = \cos \vartheta$ we can write

$$\Delta V(r, u) = \frac{\partial^2 V}{\partial r^2} + \frac{2}{r} \frac{\partial V}{\partial r} + \frac{1}{r^2} \frac{\partial}{\partial u} \left( (1 - u^2) \frac{\partial V}{\partial u} \right) = 0.$$

Solutions are the functions $V(r, u) = R(r)\Theta(u)$ with $R_n(r) = \{r^n \, ; r^{-(n+1)}\}$ and $\Theta_n(u) = \{P_n(u) \, ; Q_n(u)\}$; $n = 0, 1, 2, 3, \ldots$. The functions $P_n(u)$ and $Q_n(u)$ are the Legendre polynomials.

As the exciting potential

$$V_e = \frac{M}{4\pi\varepsilon} \frac{P_n(u)}{r^{(n+1)}} \bigg|_{n=1} = \frac{M}{4\pi\varepsilon} \frac{\cos \vartheta}{r^2} = \frac{M}{4\pi\varepsilon a^2} \left[ \left( \frac{a}{r} \right)^2 \cos \vartheta \right]$$

is a solution for $n = 1$, the overall solution is given by

$$V(r, \vartheta) = \frac{M}{4\pi\varepsilon a^2} \left\{ \begin{array}{ll} (a/r)^2 + C_1 \, r/a \\ C_2 (a/r)^2 \end{array} \right\} \cos \vartheta \; ; \quad \begin{array}{l} r \leq a \\ r \geq a \end{array} .$$

The constants $C_1$ and $C_2$ result from the boundary conditions for the potential and the normal component of the electric flux density.

$$V(r, \vartheta)\big|_{\substack{r>a \\ r \to a}} = V(r, \vartheta)\big|_{\substack{r<a \\ r \to a}} \; ; \quad \varepsilon_0 \frac{\partial V}{\partial r} \bigg|_{\substack{r>a \\ r \to a}} = \varepsilon \frac{\partial V}{\partial r} \bigg|_{\substack{r<a \\ r \to a}}$$

$$1 + C_1 = C_2 \, ; \quad C_1 - 2 = -2\,\varepsilon_0/\varepsilon \, C_2$$

$$C_2 = \frac{3}{1 + 2\,\varepsilon_0/\varepsilon} \, ; \quad C_1 = 2\frac{1 - \varepsilon_0/\varepsilon}{1 + 2\varepsilon_0/\varepsilon}$$

For the electric field it follows with $\vec{E} = -\text{grad} \, V$.

$$\vec{E} = -\vec{e}_r \frac{\partial V}{\partial r} - \frac{\vec{e}_\vartheta}{r} \frac{\partial V}{\partial \vartheta} = \vec{e}_r \, E_r + \vec{e}_\vartheta E_\vartheta$$

$$E_r = \frac{M}{4\pi\varepsilon a^3} \left\{ \begin{array}{l} 2(a/r)^3 - C_1 \\ 2C_2(a/r)^3 \end{array} \right\} \cos \vartheta \, ; \quad E_\vartheta = \frac{M}{4\pi\varepsilon a^3} \left\{ \begin{array}{l} (a/r)^3 + C_1 \\ C_2(a/r)^3 \end{array} \right\} \sin \vartheta$$

Thus the additional potential in $r < a$ describes a homogeneous $z$-directed field.

$$\frac{MC_1}{4\pi\varepsilon a^3} \left[ -\vec{e}_r \cos \vartheta + \vec{e}_\vartheta \sin \vartheta \right] = -\vec{e}_z \frac{MC_1}{4\pi\varepsilon a^3}$$

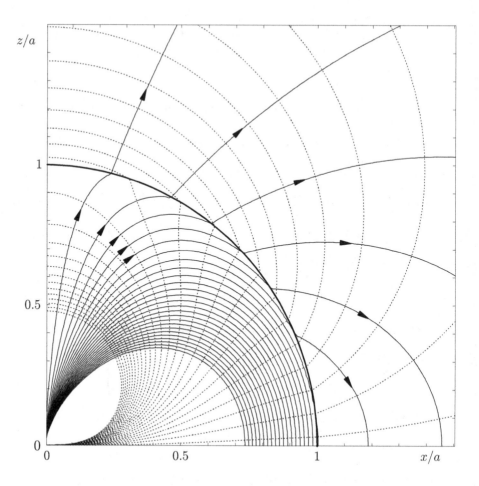

Fig. 2.7–1: Equipotential lines and electric lines of force for $\varepsilon/\varepsilon_0 = 10$

A surface charge $\sigma$ is to be placed on the sphere $r = a$, so that the field outside of $r = a$ vanishes. Therefore we have to superpose a potential in $r < a$, which is a continuous extension of the potential

$$V_c = -\frac{M}{4\pi\varepsilon a^2} C_2 (a/r)^2 \cos\vartheta \; ; \qquad r \geq a$$

in the outer space. This is the potential

$$V_c(r,\vartheta) = -\frac{M}{4\pi\varepsilon a^2} C_2 (r/a) \cos\vartheta \; ; \qquad r \leq a \, .$$

Now the potential in $r < a$ is

$$V(r,\vartheta) = \frac{M}{4\pi\varepsilon a^2} \left[ (a/r)^2 + C_1(r/a) - C_2(r/a) \right] \cos\vartheta \; ; \qquad r \leq a$$

and because of $V(r = a, \vartheta) = 0$ and $C_1 - C_2 = -1$ it follows

$$V(r, \vartheta) = \frac{M}{4\pi\varepsilon a^2} \left[ (a/r)^2 - (r/a) \right] \cos\vartheta ; \qquad r \leq a .$$

Finally the surface charge $\sigma$ is

$$\sigma = \vec{n} \left( \vec{D}_c \Big|_{\substack{r>a \\ r\to a}} - \vec{D}_c \Big|_{\substack{r<a \\ r\to a}} \right) = -\varepsilon_0 \frac{\partial V_c}{\partial r} \Big|_{\substack{r>a \\ r\to a}} + \varepsilon \frac{\partial V_c}{\partial r} \Big|_{\substack{r<a \\ r\to a}} =$$

$$= -\frac{\varepsilon + 2\varepsilon_0}{4\pi\varepsilon} \frac{M}{a^3} C_2 \cos\vartheta = -\frac{3M}{4\pi a^3} \cos\vartheta .$$

## 2.8   Potential of a Charge with Radially Dependent Density

Consider the following potential with spherical symmetry

$$V(r) = C \frac{\exp(-\alpha r)}{r} .$$

$\alpha, C$ are positive constants.

What is the charge density according to the potential in homogeneous space of permittivity $\varepsilon$?

As the spheres $r = \mathrm{const}$ are equipotential surfaces, the direction of the electric field is radial with

$$\vec{E}(r) = \vec{e}_r E(r) = -\vec{e}_r \frac{dV}{dr} = \vec{e}_r C(1 + \alpha r) \frac{\exp(-\alpha r)}{r^2} .$$

The space charge density $\varrho$ satisfies the Laplace–Poisson equation:

$$\Delta V = \frac{d^2 V}{dr^2} + \frac{2}{r} \frac{dV}{dr} = -\frac{\varrho}{\varepsilon} ; \qquad r > 0$$

$$\frac{dV}{dr} = -C(1 + \alpha r) \frac{\exp(-\alpha r)}{r^2}$$

$$\frac{d^2 V}{dr^2} = -C \left[ \alpha r + (1 + \alpha r)(-\alpha r - 2) \right] \frac{\exp(-\alpha r)}{r^3}$$

$$-\frac{\varrho}{\varepsilon} = C(\alpha r)^2 \frac{\exp(-\alpha r)}{r^3}$$

$$\varrho(r) = -\varepsilon C(\alpha r)^2 \frac{\exp(-\alpha r)}{r^3} ; \qquad r > 0 .$$

Obviously this solution is not complete, because the space charge is negative in $r > 0$ although the whole potential is positive.

The calculation of the charge $Q_k$ inside a sphere of radius $r_k$ and surface $a_k = 4\pi r_k^2$ results in

$$Q_k(r_k) = \Psi_e(r_k) = \oint_{a_k} \vec{D}\, d\vec{a} = 4\pi\varepsilon\, r_k^2\, C\,(1+\alpha r_k)\,\frac{\exp(-\alpha r_k)}{r_k^2}$$

$$\Psi_e(r_k) = 4\pi\varepsilon\, C\,(1+\alpha r_k)\,\exp(-\alpha r_k).$$

Now the limit $r_k \to 0$ leads to

$$\lim_{r_k \to 0} Q_k(r_k) = 4\pi\varepsilon\, C = Q_0\,.$$

Apparently a point charge $Q_0$ is located in the center of the coordinate system, which is not included in the calculation of the Laplace operator defined for $r > 0$.

If the charge $Q_k$ is derived by means of the integral of the incomplete solution $\varrho$ over a sphere with radius $r = r_k$ and volume $v_k$ we get

$$Q_k(r_k) = \int_{v_k} \varrho\, dv = 4\pi \int_0^{r_k} \varrho\,(r)\, r^2\, dr = -4\pi\varepsilon\, C\,\alpha^2 \int_0^{r_k} r\,\exp(-\alpha r)\, dr$$

$$= -Q_0\,\alpha^2 \left[ \frac{r\exp(-\alpha r)}{-\alpha} + \frac{1}{\alpha}\,\frac{\exp(-\alpha r)}{-\alpha} \right]_0^{r_k}$$

$$Q_k = Q_0\,[(1+\alpha r_k)\,\exp(-\alpha r_k) - 1]\,; \qquad \lim_{r_k \to 0} Q_k(r_k) = 0.$$

As the limit $r_k \to 0$ is zero in this case, the complete solution for the charge density reads

$$\varrho(r) = Q_0\,\delta(r) - \varepsilon\, C\,(\alpha r)^2\,\frac{\exp(-\alpha r)}{r^3}\,,$$

where the Dirac delta function is defined by

$$\delta(r) = 0 \quad \text{with} \quad r > 0 \quad \text{and} \quad \int_v \delta(r)\, dv = 1\,.$$

Now the charge distribution satisfies

$$\int_{v_k} \varrho(r)\, dv = \oint_{a_k} \vec{D}\, d\vec{a} = Q_k(r_k) = Q_0(1+\alpha r_k)\,\exp(-\alpha r_k)\,.$$

## 2.9 Dielectric Sphere Exposed to the Field of an Axial Line Source

A sphere with radius $r = a$ has the permittivity $\varepsilon_1$ and is surrounded by material with a permittivity $\varepsilon_2$. A homogeneous line charge of density $\lambda_{q0}$ is positioned on the $z$-axis at points $|z_q| < c < a$ inside the sphere or bounded by the interval $a < b < z_q < d$ outside the sphere.

Calculate the potential and the field for all points $r \geq 0$ respectively.

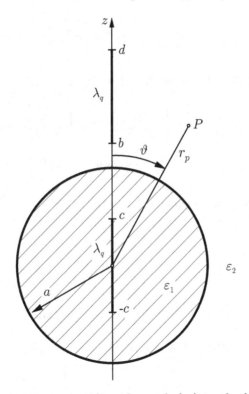

**Preconsideration:** A $z$-dependent line charge $\lambda_q(z_q)$ in the homogeneous space of permittivity $\varepsilon_0$ is defined on the interval $|z_q| < c$ on the $z$–axis. For the calculation of the potential $V(r_p, \vartheta)$ we first consider the field of a point charge $Q$ at position $z_q$. The potential at points $(r_p, \vartheta)$ is:

$$V(r_p, \vartheta) = \frac{Q}{4\pi\varepsilon_0} \frac{1}{R(z_q, r_p, \vartheta)} ; \qquad R^2 = r_p^2 + z_q^2 - 2z_q r_p \cos\vartheta$$

$$\frac{1}{R} = \frac{1}{z_q} \left[ 1 + \left(\frac{r_p}{z_q}\right)^2 - 2\left(\frac{r_p}{z_q}\right) \cos\vartheta \right]^{-\frac{1}{2}} .$$

For

$$\left| \frac{r_p}{z_q} \right|^2 + 2 \left| \frac{r_p}{z_q} \cos \vartheta \right| < 1$$

the following expansion holds.

$$\frac{1}{R} = \frac{1}{z_q} \left\{ 1 - \frac{1}{2} \left[ \left( \frac{r_p}{z_q} \right)^2 - 2 \frac{r_p}{z_q} \cos \vartheta \right] \right.$$
$$+ \frac{1}{2} \frac{3}{4} \left[ \left( \frac{r_p}{z_q} \right)^2 - 2 \frac{r_p}{z_q} \cos \vartheta \right]^2$$
$$\left. - \frac{1}{2} \frac{3}{4} \frac{5}{6} \left[ \left( \frac{r_p}{z_q} \right)^2 - 2 \frac{r_p}{z_q} \cos \vartheta \right]^3 + - \ldots \right\}$$

Rearranging in terms of $(r_p/z_q)$ leads to

$$\frac{1}{R} = \frac{1}{z_q} \left\{ \left[ 1 \right] + \frac{r_p}{z_q} \left[ \cos \vartheta \right] + \left( \frac{r_p}{z_q} \right)^2 \left[ -\frac{1}{2} + \frac{3}{2} \cos^2 \vartheta \right] + \right.$$
$$\left. + \left( \frac{r_p}{z_q} \right)^3 \left[ -\frac{3}{2} \cos \vartheta + \frac{5}{2} \cos^3 \vartheta \right] + \left( \frac{r_p}{z_q} \right)^4 \left[ \ldots \right] + \ldots \right\}.$$

The expressions in squared brackets are called Legendre polynomials

$$P_0 = 1,$$
$$P_1 = \cos \vartheta,$$
$$P_2 = \frac{1}{2} \left( 3 \cos^2 \vartheta - 1 \right),$$
$$P_3 = \frac{\cos \vartheta}{2} \left( 5 \cos^2 \vartheta - 3 \right) \ldots.$$

They satisfy the orthogonality relation

$$\int_{-1}^{+1} P_n(u) P_k(u) du = \begin{cases} 0 & n \neq k \\ \dfrac{2}{2n+1} & n = k \end{cases}.$$

With the Legendre polynomials the reciprocal distance $1/R$ takes the form

$$\frac{1}{R} = \sum_{n=0}^{\infty} \left\{ \begin{array}{l} \dfrac{1}{z_q} \left( \dfrac{r_p}{z_q} \right)^n \\[2mm] \dfrac{1}{r_p} \left( \dfrac{z_q}{r_p} \right)^n \end{array} \right\} P_n(\cos \vartheta) \, ; \qquad \begin{array}{l} r_p \leq z_q \\[4mm] z_q \leq r_p, \end{array}$$

where an analog expansion for $z_q \leq r_p$ has been used.

For the interpretation of the reciprocal distance we can employ a Taylor expansion around the point $z_q = 0$:

$$\frac{1}{R} = \left[ r_p^2 + z_q^2 - 2 r_p z_q \cos \vartheta \right]^{-1/2} = f(z_q)$$

$$= f(0) + f'(z_q) \Big|_{z_q=0} z_q + \frac{f''(z_q)}{2!} \Big|_{z_q=0} z_q^2 + \dots$$

For cartesian coordinates applies:

$$\frac{1}{R} = \left[ x_p^2 + y_p^2 + (z_p - z_q)^2 \right]^{-1/2} \rightarrow \frac{\partial}{\partial z_q} = -\frac{\partial}{\partial z_p}$$

$$\frac{\partial^n f(z_q, z_p)}{\partial z_q^n} \Big|_{z_q=0} = (-1)^n \frac{\partial^n}{\partial z_p^n} f(z_q, z_p) \Big|_{z_q=0} \quad ; \quad z_p = r_p \cos \vartheta.$$

Thus:

$$\frac{1}{R} = \frac{1}{r_p} - z_q \frac{\partial}{\partial z_p} \left( \frac{1}{r_p} \right) + \frac{z_q^2}{2!} \frac{\partial^2}{\partial z_p^2} \left( \frac{1}{r_p} \right) - \dots + z_q^n \frac{(-1)^n}{n!} \frac{\partial^n}{\partial z_p^n} \left( \frac{1}{r_p} \right) + \dots$$

The comparison of the expansions leads to the fundamental relation

$$\frac{P_n(\cos \vartheta)}{r_p^{n+1}} = \frac{(-1)^n}{n!} \frac{\partial^n}{\partial z_p^n} \left( \frac{1}{r_p} \right).$$

This result can be used to express the potential of the line charge $\lambda_q(z_q)$ with $|z_q| < c$ in the following way:

$$V(r_p, \vartheta) = \frac{1}{4\pi\varepsilon_0} \int_{-c}^{c} \frac{\lambda_q(z_q)}{R(z_q, r_p, \vartheta)} dz_q$$

$$r_p > c: \qquad V(r_p, \vartheta) = \frac{1}{4\pi\varepsilon_0} \int_{-c}^{c} \underbrace{\frac{1}{r_p} \sum_{n=0}^{\infty} \left( \frac{z_q}{r_p} \right)^n P_n(\cos \vartheta)}_{\dfrac{1}{R(z_q, r_p, \vartheta)}} \lambda_q(z_q) dz_q.$$

Now if we evaluate the integral for the first term of the series we get

$$V_0(r_p) = \frac{1}{4\pi\varepsilon_0} \frac{1}{r_p} \int_{-c}^{c} \lambda_q(z_q) dz_q = \frac{Q}{4\pi\varepsilon} \frac{1}{r_p}$$

$$\text{with} \quad Q = \int_{-c}^{c} \lambda_q(z_q) dz_q.$$

This is the result for the potential $V_0(r_p)$ of the total charge $Q$ located at the center of the coordinate system. The expression for $n = 1$ yields

$$V_1(r_p, \vartheta) \quad = \quad \frac{1}{4\pi\varepsilon_0} \int_{-c}^{c} \frac{z_q}{r_p^2} \lambda_q(z_q) P_1(\cos\vartheta) dz_q \quad = \quad \frac{m_1}{4\pi\varepsilon_0} \frac{\cos\vartheta}{r_p^2}$$

$$\text{with} \quad m_1 \quad = \quad \int_{-c}^{c} z_q \lambda(z_q) dz_q$$

and describes the potential of a dipole moment $m_1$. The series terms with $n \geq 2$ define axial multipoles of higher order

$$m_n \quad = \quad \int_{-c}^{c} z_q^n \lambda_q(z_q) dz_q .$$

Their potentials

$$V_n \quad = \quad \frac{m_n}{4\pi\varepsilon_0} \frac{P_n(\cos\vartheta)}{r_p^{n+1}}$$

are in total

$$V \quad = \quad \sum_{n=0}^{\infty} V_n$$

the potential of the line charge.

Example: Homogeneous line charge:

$$\lambda_{q0} \quad = \quad Q/2c$$

$$m_n \quad = \quad \frac{Q}{2c} \int_{-c}^{c} z_q^n dz_q \quad = \quad \frac{Q}{2c} \frac{c^{n+1} - (-c)^{n+1}}{n+1}$$

$$m_{2k+1} \quad = \quad 0; \qquad m_{2k} \quad = \quad \frac{Qc^{2k}}{2k+1}; \qquad k = 0, 1, 2, \ldots$$

$$\lim_{c \to 0} m_{2k} \quad = \quad \begin{cases} 0 & k > 0 \\ Q & k = 0 \end{cases}$$

The initial problems are now solved by means of the expansion of the reciprocal distance.

## Charge within the Sphere

A homogeneous line charge $\lambda_{q0}$ of range $|z_q| < c < a$ is given in the domain $r < a$ inside the sphere.

The exciting potential $V_e(r, \vartheta)$ of the line charge in homogeneous space of permittivity $\varepsilon_1$ is

$$V_e(r, \vartheta) = \sum_{n=0}^{\infty} V_n = \sum_{n=0}^{\infty} \frac{m_n}{4\pi\varepsilon_1} \frac{P_n(\cos \vartheta)}{r^{n+1}} ; \qquad r \geq c .$$

An analog expression holds for $r \leq c$, but is not directly needed for the following calculation.

$$m_{2k} = 2\lambda_{q0} \frac{c^{2k+1}}{2k+1} ; \qquad m_{2k+1} = 0$$

$$V_e(r, \vartheta) = \frac{\lambda_{q0}}{2\pi\varepsilon_1} \sum_{k=0}^{\infty} \frac{1}{2k+1} \left(\frac{c}{r}\right)^{2k+1} P_{2k}(\cos \vartheta) ; \qquad r \geq c$$

The resulting potential is:

$$V = V_e(r, \vartheta) + V_s(r, \vartheta) ; \qquad r \geq c .$$

In $r \geq c$ applies with the constants $A_k$

$$V = \frac{\lambda_{q0}}{2\pi\varepsilon_1} \sum_{k=0}^{\infty} \frac{1}{2k+1} \left[ \left(\frac{c}{r}\right)^{2k+1} + A_k \begin{cases} \left(\dfrac{r}{a}\right)^{2k} & c \leq r \leq a \\[2mm] \left(\dfrac{a}{r}\right)^{2k+1} & r \geq a . \end{cases} \right] P_{2k}(\cos \vartheta);$$

This expression already satisfies the continuity of the potential in $r = a$. The constants $A_k$ are determined by evaluating the boundary condition for the normal component of the electric flux density.

$$\left( \varepsilon_1 \frac{\partial V}{\partial r} \bigg|_{r<a} - \varepsilon_2 \frac{\partial V}{\partial r} \bigg|_{r>a} \right) \bigg|_{r \to a} = 0$$

$$\left[ -\frac{2k+1}{c} \left(\frac{c}{a}\right)^{2k+2} + A_k 2k\frac{1}{a} \right] = \frac{\varepsilon_2}{\varepsilon_1} \left[ -\frac{2k+1}{c} \left(\frac{c}{a}\right)^{2k+2} - A_k \frac{2k+1}{a} \right]$$

$$\Rightarrow \quad A_k = \frac{(2k+1)(1 - \varepsilon_2/\varepsilon_1)}{2k + \varepsilon_2/\varepsilon_1 (2k+1)} \left(\frac{c}{a}\right)^{2k+1}$$

In $r > a$ the potential is independent of $a$. The limit case $\varepsilon_1 = \varepsilon_2$ provides directly the sole exciting potential. For $\varepsilon_2 \to \infty$ it follows

$$A_k|_{\varepsilon_2 \to \infty} = -\left(\frac{c}{a}\right)^{2k+1} ; \qquad c \leq r \leq a$$

$$V\big|_{\varepsilon_2\to\infty} = \frac{\lambda_{q0}}{2\pi\varepsilon_1} \sum_{k=0}^{\infty} \frac{1}{2k+1} \left[ \left(\frac{c}{r}\right)^{2k+1} - \underbrace{\left(\frac{c}{a}\right)^{2k+1} \left(\frac{r}{a}\right)^{2k}}_{a/r\,(cr/a^2)^{2k+1}} \right] P_{2k}(\cos\vartheta)$$

$$= V_e\,(r,\vartheta) - \frac{a}{r}\,V_e\left(a^2/r,\vartheta\right) = V_e(\vec{r}) - \frac{a}{r}\,V_e(\vec{r}(a/r)^2).$$

The resulting potential is derived directly from the exciting potential (Law of potential mirroring).

## Charge outside the Sphere

A homogeneous line charge $\lambda_{q0}$ is positioned in front of the dielectric sphere of permittivity $\varepsilon_1$ on the $z$–axis with $a < b < z_q < d$.

The exciting potential $V_e(r,\vartheta)$ of the line charge in a homogeneous space of permittivity $\varepsilon_2$ is

$$V_e(r,\vartheta) = \frac{1}{4\pi\varepsilon_2} \int_b^d \frac{\lambda_{q0}}{R(z_q,r,\vartheta)} dz_q \;; \qquad R^2 = r^2 + z_q^2 - 2r\,z_q\,\cos\vartheta$$

$$V_e(r,\vartheta) = \frac{\lambda_{q0}}{4\pi\varepsilon_2} \int_b^d \sum_{n=0}^{\infty} \left\{ \begin{array}{l} \dfrac{1}{z_q}\left(\dfrac{r}{z_q}\right)^n \\[2mm] \dfrac{1}{r}\left(\dfrac{z_q}{r}\right)^n \end{array} \right\} P_n(\cos\vartheta)\,dz_q \;; \qquad \begin{array}{l} r \le z_q \\[4mm] r \ge z_q \end{array}\quad.$$

Because the position of the line charge has moved outside the center of the coordinate system, we obtain three integrals.

$$r \le b: \quad V_e = \frac{\lambda_{q0}}{4\pi\varepsilon_2} \sum_{n=0}^{\infty} r^n\,P_n(\cos\vartheta) \int_b^d z_q^{-(n+1)}dz_q$$

$$= \frac{\lambda_{q0}}{4\pi\varepsilon_2} \sum_{n=0}^{\infty} \frac{1}{n}\left[1 - \left(\frac{b}{d}\right)^n\right]\left(\frac{r}{b}\right)^n P_n(\cos\vartheta)$$

$$b \le r \le d: \quad V_e = \frac{\lambda_{q0}}{4\pi\varepsilon_2} \sum_{n=0}^{\infty} \left[ r^{-(n+1)} \int_b^r z_q^n dz_q + r^n \int_r^d z_q^{-(n+1)}dz_q \right] P_n(\cos\vartheta)$$

$$= \frac{\lambda_{q0}}{4\pi\varepsilon_2} \sum_{n=0}^{\infty} \left[ \frac{1}{n+1}\left[1 - \left(\frac{b}{r}\right)^{n+1}\right] + \frac{1}{n}\left[1 - \left(\frac{r}{d}\right)^n\right] \right] P_n(\cos\vartheta)$$

$$r \ge d: \quad V_e = \frac{\lambda_{q0}}{4\pi\varepsilon_2} \sum_{n=0}^{\infty} r^{-(n+1)} \int_b^d z_q^n\,dz_q P_n(\cos\vartheta)$$

$$= \frac{\lambda_{q0}}{4\pi\varepsilon_2} \sum_{n=0}^{\infty} \frac{1}{n+1}\left[1 - \left(\frac{b}{d}\right)^{n+1}\right]\left(\frac{d}{r}\right)^{n+1} P_n(\cos\vartheta)$$

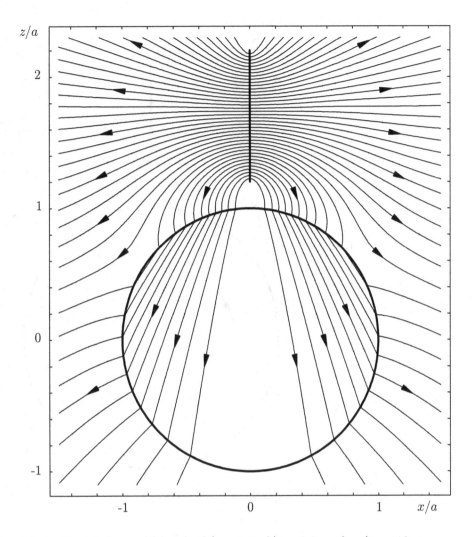

Fig. 2.9–1: Electric lines of force for $b/a = 1.2$, $d/a = 2.2$, and $\varepsilon_1/\varepsilon_2 = 10$

The resulting potential is in total

$$V(r, \vartheta) = V_e + V_s = V_e + \frac{\lambda_{q_0}}{4\pi\varepsilon_2} \sum_{n=0}^{\infty} B_n \left\{ \begin{array}{l} (r/a)^n \\ (a/r)^{n+1} \end{array} \right\} P_n(\cos\vartheta); \quad \begin{array}{l} r \leq a \\ r \geq a \end{array} .$$

As before, this expression already satisfies the continuity of the potential in $r = a$.

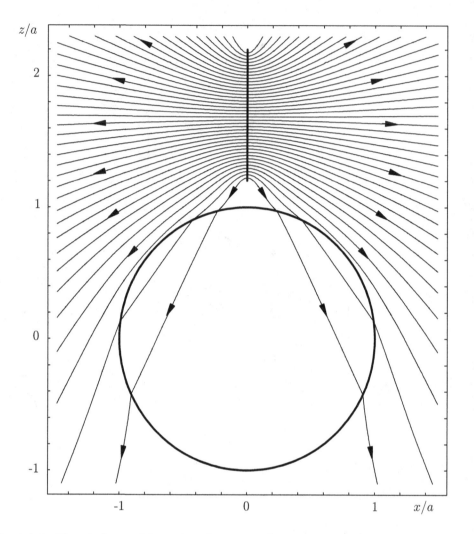

Fig. 2.9–2: Electric lines of force for $b/a = 1.2$, $d/a = 2.2$, and $\varepsilon_1/\varepsilon_2 = 0.5$

The constants $B_n$ follow from the boundary condition

$$\left( \varepsilon_1 \left. \frac{\partial V}{\partial r} \right|_{r<a} - \varepsilon_2 \left. \frac{\partial V}{\partial r} \right|_{r>a} \right)\Bigg|_{r\to a} = 0$$

with the result

$$B_n = \frac{1 - \varepsilon_1/\varepsilon_2}{1 + n(1 + \varepsilon_1/\varepsilon_2)} \left( \frac{a}{b} \right)^n \left[ 1 - \left( \frac{b}{d} \right)^n \right] .$$

Once again the limit case $\varepsilon_1 = \varepsilon_2$ brings out the exciting potential. For large $\varepsilon_1$ we get

$$B_n\big|_{\varepsilon_1 \to \infty} = -\frac{1}{n} \left(\frac{a}{b}\right)^n \left[1 - \left(\frac{b}{d}\right)^n\right]$$

and

$$V\big|_{\varepsilon_1 \to \infty\,;\,r \geq a} = V_e - \frac{\lambda_{q0}}{4\pi\varepsilon_2} \sum_{n=0}^{\infty} \underbrace{\frac{1}{n}\left(\frac{a}{b}\right)^n \left[1 - \left(\frac{b}{d}\right)^n\right] \left(\frac{a}{r}\right)^{n+1}}_{\displaystyle \frac{1}{n}\frac{a}{r}\left(\frac{a^2/r}{b}\right)^n \left[1 - \left(\frac{b}{d}\right)^n\right]} P_n(\cos\vartheta)$$

$$= V_e(r,\vartheta) - \frac{a}{r} V_e(a^2/r,\vartheta) = V_e(\vec{r}) - \frac{a}{r} V_e(\vec{r}(a/r)^2)\,.$$

Thus again the resulting potential is derived directly from the exciting potential.


## 2.10   Concentric Cylinders With Given Potential

Two concentric cylinders with radii $\varrho = a$ and $\varrho = b > a$ are on the potentials $V(\varrho = a, z) = V_0 \sin(\pi z/h)$ and $V(\varrho = b, z) = V_0 \sin(2\pi z/h)$. The planes $z = 0$ and $z = h$ are equipotential surfaces $V(\varrho, z = 0) = 0$ and $V(\varrho, z = h) = 0$. The permittivity $\varepsilon$ is constant.

Calculate the potential $V(\varrho, z)$ in the domain $(a \leq \varrho \leq b; 0 \leq z \leq h)$ and determine the charge density $\sigma(\varrho)$ in the (conducting) plane $z = 0$ within $a < \varrho < b$.

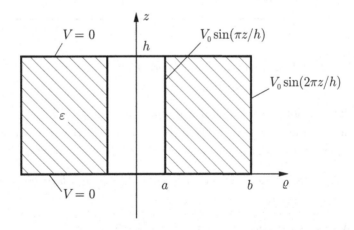

Because the specified potentials on the cylinders $\varrho = \text{const}$ depend on the $z$-coordinate, we have to choose orthogonal functions in this coordinate, e.g. trigonometric functions

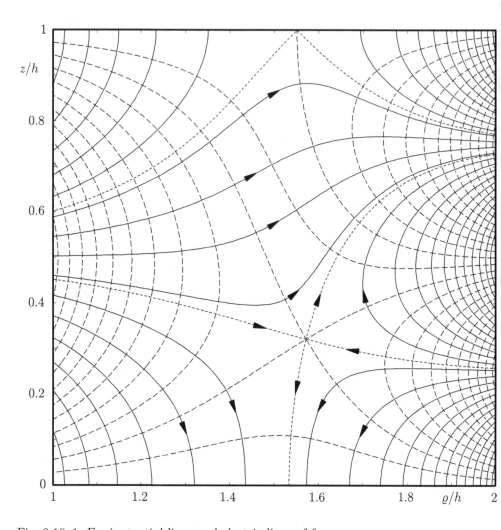

Fig. 2.10–1: Equipotential lines and electric lines of force

$\sin(mz)$, for the solution of the Laplace equation $\Delta V(\varrho, z) = 0$. The dependence on the coordinate $\varrho$ is then given by the modified Bessel functions $I_0(m\varrho)$ and $K_0(m\varrho)$, which are of order zero because of the rotational symmetry.

The given potential on the surfaces $\varrho = a$ and $\varrho = b$ requires the parameters $m = \pi/h$ and $m = 2\pi/h$.

With this parameters and the linear combination

$$R_n(\varrho, \eta) = I_0(n\pi\varrho/h)\, K_0(n\pi\eta/h) - K_0(n\pi\varrho/h)\, I_0(n\pi\eta/h),$$

that obeys the property $R_n(\eta, \eta) = 0$, the solution for the potential reads

$$V(\varrho, z) = V_0 \left( \sin(\pi z/h) \frac{R_1(\varrho, b)}{R_1(a, b)} + \sin(2\pi z/h) \frac{R_2(\varrho, a)}{R_2(b, a)} \right); \quad \begin{array}{l} a \le \varrho \le b \\ 0 \le z \le h. \end{array}$$

For the surface charge density $\sigma(\varrho)$ on the boundary $z = 0$ it follows

$$\sigma(\varrho) = -\varepsilon \left. \frac{\partial V}{\partial z} \right|_{z=0} = -\varepsilon V_0 \, \pi/h \left( \frac{R_1(\varrho, b)}{R_1(a, b)} + 2 \frac{R_2(\varrho, a)}{R_2(b, a)} \right); \quad a < \varrho < b.$$

## 2.11    Method of Images For Conducting Spheres

The center of a spherical conducting surface of radius $a$ and potential $V_0$ lies in the center of the coordinate system, where in addition a point charge $Q$ is positioned. Within the plane $z = c > a$ a circular surface charge $\sigma$ of radius $b$ is centered around the $z$-axis. The permittivity $\varepsilon$ is constant.

Calculate the potential on the $z$-axis. What is the total charge of the sphere, when the potential is $V_0 = 0$?

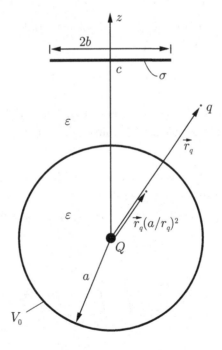

The fields inside and outside the conducting sphere are determined independently.

Inside the sphere $|\vec{r}| \leq a$ the potential is the sum of a constant potential and the potential of a point charge. As the potential on the surface $|\vec{r}| = a$ is $V_0$, the solution is

$$V(r) = V_0 + \frac{Q}{4\pi\varepsilon}\left(\frac{1}{r} - \frac{1}{a}\right); \qquad r \leq a$$

and on the $z$-axis applies

$$V(z) = V_0 + \frac{Q}{4\pi\varepsilon}\left(\frac{1}{|z|} - \frac{1}{a}\right); \qquad |z| \leq a.$$

For the determination of the potential outside the sphere the method of images for conducting spheres is used. However, the mirrored surface charge is inhomogeneously distributed on a curvilinear surface. Its contribution to the potential leads to a complicated integration. A considerably easier approach is to derive this contribution by evaluating the expression for the potential of the initial charge at mirrored observation points. Thus no additional calculation is needed (cp. problem 2.9).

For example, if a point charge $Q$ is located in front of a grounded sphere of radius $a$ at position $\vec{r}_q$ with $|\vec{r}_q| > a$, then the resulting potential $V(\vec{r}_p)$ at points $\vec{r}_p$ with $|\vec{r}_p| \geq a$ is the sum of the potential $V_e(\vec{r}_p)$ of the charge $Q$ in homogeneous space and of the potential of the mirrored charge $-Qa/|\vec{r}_q|$ at position $\vec{r}_q(a/r_q)^2$.

We get the same result if the total potential $V(\vec{r}_p)$ is derived only by means of the exciting potential.

$$V(\vec{r}_p) = V_e(\vec{r}_p) - \frac{a}{r_p}V_e(\vec{r}_p(a/r_p)^2); \qquad |\vec{r}_p| > a$$

This approach is even independent of the initial charge distribution, and hence it is preferable, if the charge distribution does not comprise simple point charges. This is actually the case.

In order to apply this approach for the determination of the potential on the rotational $z$-axis of the surface charge in front of the conducting sphere, we first have to calculate the exciting potential $V_e(z)$ of the circular surface charge in the homogeneous space.

$$\begin{aligned}
V_e(z) &= \frac{1}{4\pi\varepsilon}\int\frac{\sigma(\vec{r}_q)}{|\vec{r}_p - \vec{r}_q|}\,da = \\[2mm]
&= \frac{\sigma}{4\pi\varepsilon}\int_{\varrho_q=0}^{b}\int_{\varphi_q=0}^{2\pi}\frac{\varrho_q\,d\varphi_q\,d\varrho_q}{\sqrt{\varrho_q^2 + (z-c)^2}} = \frac{\sigma}{2\varepsilon}\int_{\varrho_q=0}^{b}\frac{\varrho_q\,d\varrho_q}{\sqrt{\varrho_q^2 + (z-c)^2}} \\[2mm]
&= \frac{\sigma}{2\varepsilon}\sqrt{\varrho_q^2 + (z-c)^2}\Big|_0^b = \frac{\sigma}{2\varepsilon}\left(\sqrt{b^2 + (z-c)^2} - |z - c|\right)
\end{aligned}$$

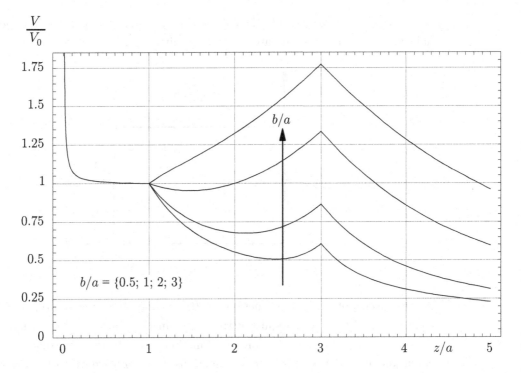

Fig. 2.11–1: Potential on the $z$-axis for $Q = 10^{-12}$ [As], $\varepsilon = \varepsilon_0$, $\sigma = 10^{-11}$[As/m$^2$], and $c/a = 3$

Now with the presented method of images (for potentials) the resulting potential of our problem is

$$V(z) = V_e(z) - \frac{a}{|z|} V_e(a^2/z) + \frac{a}{|z|} V_0 ; \qquad |z| \geq a .$$

The last term accounts for the given potential $V_0$ on the surface of the sphere. This term vanishes, i.e. $V(a) = V_0 = 0$, if the additional total charge of $Q_z = -4\pi\varepsilon a V_0$ is brought onto the spherical surface.

## 2.12 Rectangular Cylinder with Given Potential

The potentials on the boundary of a cylinder with rectangular cross section are

$$V(x = 0, y) = V(x = a, y) = 0 ; \qquad V(y = 0, x) = V_2 \cos(\pi x/a)$$

$$V(y = b, x) = V_1 \sin(\pi x/a) .$$

Calculate the potential inside the cylinder $(0 \leq x \leq a\,; 0 \leq y \leq b)$. The potential is independent of the $z$-coordinate and the permittivity $\varepsilon$ is constant.

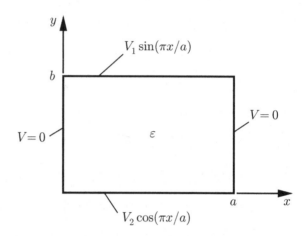

As the given potentials are functions of the $x$-coordinate, the solutions of Laplace's equation $\Delta V(x,y) = 0$ require orthogonal functions for the description of the $x$-dependency, accordingly trigonometric functions. Due to the homogeneous boundary conditions in $x = 0$ and $x = a$ the functions $\sin(m\pi x/a)$ have to be considered. In consequence the dependency of the $y$-coordinate is given by hyperbolic or exponential functions. With these functions a possible ansatz for the potential inside the rectangle $(0 \leq x \leq a; 0 \leq y \leq b)$ is

$$V(x,y) = \sum_{m=1}^{\infty} [A_m \sinh(m\pi y/a) + B_m \cosh(m\pi y/a)] \sin(m\pi x/a).$$

It is useful to split the problem, because of the simple description of the boundary values. For the first subproblem with solution $V_I$ only the boundary values at $y = b$ exist, whereas in the second subproblem with solution $V_{II}$ only the boundary values at $y = 0$ are considered. All other boundaries have the potential $V = 0$, respectively. The solution of the original problem is $V = V_I + V_{II}$.

The potential on the surface $y = b$ suggests, that only the term with $m = 1$ is needed in the summation above for the potential $V_I$. With the requirement $V_I(x, y = 0) = 0$ we get the solution

$$V_I(x,y) = V_1 \sin(\pi x/a) \frac{\sinh(\pi y/a)}{\sinh(\pi b/a)}.$$

Things are different for the second subproblem with solution $V_{II}$. With the given potential on the surface $y = 0$ the potential is discontinuous at $x = 0$ and $x = a$. As a

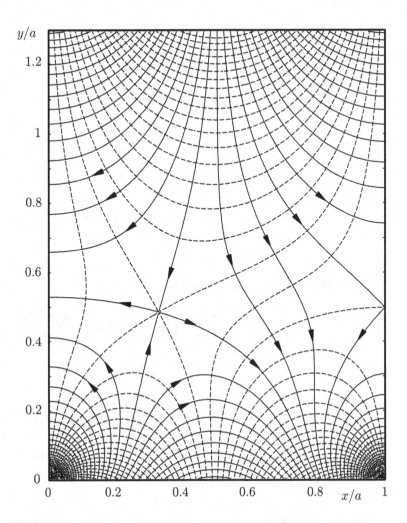

Fig. 2.12–1: Equipotential lines and electric lines of force for $V_1 = V_2$

consequence, the following complete approach has to be considered.

$$V_{II}(x, y) = \sum_{m=1}^{\infty} C_m \sin(m\pi x/a) \frac{\sinh(m\pi(b-y)/a)}{\sinh(m\pi b/a)} \, .$$

Here the arguments of the hyperbolic functions have been adapted to satisfy the boundary condition $V_{II}(x, y = b) = 0$. The division by the hyperbolic function with constant argument simplifies the evaluation of the boundary condition and thus leads to

$$V_{II}(x, y = 0) = \sum_{m=1}^{\infty} C_m \sin(m\pi x/a) = V_2 \cos(\pi x/a) \, .$$

Now, making use of the orthogonality relations for trigonometric functions by multiplication with $\sin(n\pi x/a)$ and integration from $x = 0$ to $x = a$ results in

$$\sum_m C_m \int_0^a \sin(m\pi x/a) \sin(n\pi x/a)\, dx = V_2 \int_0^a \cos(\pi x/a) \sin(n\pi x/a)\, dx$$

$$C_n a/2 = -V_2 \left[ \frac{\cos((n+1)\pi x/a)}{2(n+1)\pi/a} + \frac{\cos((n-1)\pi x/a)}{2(n-1)\pi/a} \right]_0^a$$

$$= V_2 \frac{a}{\pi} \frac{4k}{4k^2 - 1}\ ; \qquad n = 2k\ ; \quad k = 1, 2, \ldots$$

$$\Rightarrow C_{2k} = V_2 \frac{8}{\pi} \frac{k}{4k^2 - 1}\ .$$

Finally the solution of the subproblem is

$$V_{II}(x, y) = V_2 \frac{8}{\pi} \sum_{k=1}^\infty \frac{k}{4k^2 - 1} \sin(2k\pi x/a) \frac{\sinh(2k\pi(b - y)/a)}{\sinh(2k\pi b/a)}$$

and the total solution is    $V(x, y) = V_I(x, y) + V_{II}(x, y)\,.$

## 2.13    Potential of Hemispherical Charge Distributions

A sphere of radius $a$ is intersected by the plane $z = 0$ running through its center. The two hemispheres $z > 0$ and $z < 0$ carry the space charge densities $\pm\varrho$. The permittivity $\varepsilon$ is constant.

Calculate the potential and the field on the $z$–axis. Simplify the expression for the field for the case $z \gg a$ and for $z = 0$.

The determination of the potential on the $z$-axis requires the solution of

$$V(z) = \frac{1}{4\pi\varepsilon} \int \frac{\varrho(\vec{r}_q)}{|\vec{e}_z z - \vec{r}_q|}\, dv\ ; \quad V(-z) = -V(z)\ ; \quad |\vec{e}_z z - \vec{r}_q| = r\,.$$

With the reciprocal distance (cp. problem 2.9)

$$\frac{1}{r} = \sum_{n=0}^\infty \left\{ \begin{array}{l} \dfrac{1}{|z|} \left(\dfrac{r_q}{|z|}\right)^n \\[2ex] \dfrac{1}{r_q} \left(\dfrac{|z|}{r_q}\right)^n \end{array} \right\} P_n\left(\cos\vartheta_q\right)\ ; \qquad \begin{array}{l} |z| \geq r_q \\[2ex] |z| \leq r_q \end{array}$$

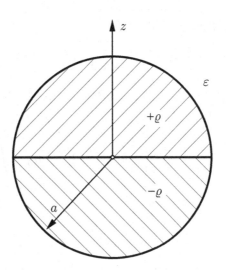

the potential for $z > a$ becomes

$$V(z) = \frac{1}{4\pi\varepsilon} \int\limits_{r_q=0}^{a} \int\limits_{\vartheta_q=0}^{\pi/2} \int\limits_{\varphi_q=0}^{2\pi} \frac{\varrho}{z} \sum_{n=0}^{\infty} \left(\frac{r_q}{z}\right)^n [P_n(\cos\vartheta_q) +$$

$$- P_n(\cos(\pi - \vartheta_q))] \, r_q^2 \sin\vartheta_q \, d\varphi_q \, d\vartheta_q \, dr_q$$

$$= \frac{1}{2\varepsilon} \frac{\varrho}{z} \sum_{n=0}^{\infty} \frac{1}{z^n} \frac{a^{n+3}}{n+3} \int\limits_{u=0}^{1} [P_n(u) - P_n(-u)] \, du \, ; \qquad u = \cos\vartheta_q \, .$$

With $\quad P_n(-u) = (-1)^n P_n(u) \quad$ and $\quad \displaystyle\int P_n(u)du = \frac{P_{n+1} - P_{n-1}}{2n+1} \quad$ it follows

$$V(z) = \frac{\varrho a^2}{2\varepsilon} \sum_{k=0}^{\infty} \frac{1}{k+2} \left(\frac{a}{z}\right)^{2(k+1)} \left.\frac{P_{2k+2} - P_{2k}}{4k+3}\right|_0^1 \, ; \qquad n = 2k+1 \, .$$

The Legendre polynomials take the values

$$P_{2k}(1) = 1 \quad \text{and} \quad P_{2k}(0) = (-1)^k \frac{1\cdot 3\cdot 5\ldots(2k-1)}{2\cdot 4\cdot 6\ldots 2k}$$

$$\Rightarrow \quad \left.\frac{P_{2k+2} - P_{2k}}{4k+3}\right|_0^1 = P_{2k}(0) \frac{\frac{2k+1}{2k+2}+1}{4k+3} = \frac{P_{2k}(0)}{2(k+1)},$$

thus

$$V(z) = \frac{\varrho a^2}{4\varepsilon} \sum_{k=0}^{\infty} \frac{1}{(k+1)(k+2)} \left(\frac{a}{z}\right)^{2(k+1)} P_{2k}(0) \, ; \quad z > a \, ; \quad V(-z) = -V(z) \, .$$

The electric field becomes

$$E_z = -\frac{\partial V}{\partial z} = \frac{\varrho a}{2\varepsilon} \sum_{k=0}^{\infty} \frac{1}{k+2} \left(\frac{a}{z}\right)^{2k+3} P_{2k}(0); \quad z > a; \quad E(z) = E(-z).$$

For points $0 < z < a$ the integration has to be split up because of the expansion of the reciprocal distance $1/r$.

$$V(z) = \frac{\varrho}{2\varepsilon} \sum_{n=0}^{\infty} \left[ \int_{r_q=0}^{z} \frac{1}{z} \left(\frac{r_q}{z}\right)^n r_q^2 \, dr_q + \int_{z}^{a} \frac{1}{r_q} \left(\frac{z}{r_q}\right)^n r_q^2 \, dr_q \right] \cdot$$

$$\cdot \int_{0}^{1} [P_n(u) - P_n(-u)] \, du$$

$$= \frac{\varrho}{2\varepsilon} \sum_{k=0}^{\infty} \left[ \frac{z^2}{2(k+2)} + \frac{a^2}{2k-1} \left( \left(\frac{z}{a}\right)^2 - \left(\frac{z}{a}\right)^{2k+1} \right) \right] \frac{P_{2k}(0)}{(k+1)}$$

$$V(z) = \frac{\varrho a^2}{2\varepsilon} \sum_{k=0}^{\infty} \left[ \frac{4k+3}{2(k+2)} \left(\frac{z}{a}\right)^2 - \left(\frac{z}{a}\right)^{2k+1} \right] \frac{P_{2k}(0)}{(k+1)(2k-1)}$$

For the electric field the expression

$$E_z = -\frac{\varrho a}{2\varepsilon} \sum_{k=0}^{\infty} \left[ \frac{4k+3}{k+2} \frac{z}{a} - (2k+1) \left(\frac{z}{a}\right)^{2k} \right] \frac{P_{2k}(0)}{(k+1)(2k-1)}; \quad 0 < z < a$$

holds, which is a continuous function for $z \to a$, thus takes the value of $E_z$, that has been derived from the potential in $z > a$.

The expression for the field simplifies for $z \to 0$ to

$$E_z(z = 0) = -\frac{\varrho a}{2\varepsilon},$$

because only the series term $k = 0$ contributes to the field. This result can also be obtained by a direct integration for the point $z = 0$.

$$\vec{E} = \frac{1}{4\pi\varepsilon} \int \varrho \frac{\vec{r}}{r^3} \, dv = \vec{e}_z \, E_z; \qquad \vec{r} = -\vec{r}_q$$

$$E_z = \frac{1}{4\pi\varepsilon} \int \varrho \frac{\vec{e}_z(-\vec{r}_q)}{r_q^3} \, dv; \qquad dv = r_q^2 \sin\vartheta_q \, dr_q \, d\vartheta_q d\varphi_q$$

$$= -\frac{\varrho}{2\varepsilon} \int_{r_q=0}^{a} \left[ -\int_{-1}^{0} u \, du + \int_{0}^{1} u \, du \right] dr_q; \qquad u = \cos\vartheta_q$$

$$E_z = -\frac{\varrho a}{2\varepsilon}$$

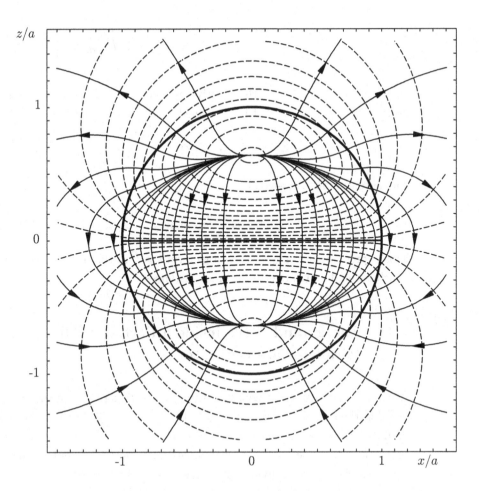

Fig. 2.13–1: Equipotential lines and electric lines of force inside and outside of the charge distribution (In the charged area the flux inside a "flux tube" is no longer constant like in the charge-free space, but decreases until the electric field is zero.)

For great distances $z \gg |a|$ it is possible to calculate the potential by means of the positive and negative charge center. The (positive) charge center of the hemispherical charge in $z > 0$ is ($Q$ is the total charge):

$$\vec{r}_{qs} = \frac{1}{Q} \int \vec{r}_q \, \varrho \, dv = \vec{e}_z \, r_{qs} \, ; \qquad r_{qs} = \frac{3}{a^3} \int\limits_{\vartheta_q=0}^{\pi/2} \int\limits_{r_q=0}^{a} \vec{e}_z \, \vec{r}_q \, r_q^2 \, dr_q \, \sin \vartheta_q d\vartheta_q$$

$$u = \cos \vartheta_q \; ; \qquad r_{qs} = \frac{3}{4}a \int_0^1 u\,du = \frac{3}{8}a \; .$$

Hence the distance of the the positive and negative charge center is $s = \frac{6}{8}a$ and for points $z \gg |a|$ the potential $V_s$ is defined by the dipole moment

$$\vec{m} = \vec{e}_z\, m \; ; \qquad m = Q \cdot s = \varrho\,\frac{4}{6}\pi a^3 \cdot \frac{6}{8}a = \frac{\pi}{2}\varrho\, a^4 .$$

The potential on the $z$-axis is

$$V_s = \frac{\varrho\, a^4}{8\varepsilon\, z^2}\, \mathrm{sign}(z) \; ; \qquad |z| \gg a .$$

This expression agrees with the series term $k = 0$ in the series expansion of $V(z \gg a)$, which is the dominant term for great distances.

## 2.14 Energy and Force inside a Partially Filled Parallel-Plate Capacitor

A parallel-plate capacitor consists of two conducting plates on the surfaces $y = 0$ and $y = d$ in the domain $0 < x < b \gg d$ and $0 < z < b$. Within the homogeneous space of permittivity $\varepsilon_0$ its capacitance is $C_0$. The capacitor has been charged to the voltage $U_0$, but is now disconnected from the source and a dielectric has been partially inserted, so that a plane $x = x_p$ divides the areas of different materials.

Calculate the charge distributions on the plates. What is the force on the dielectric?

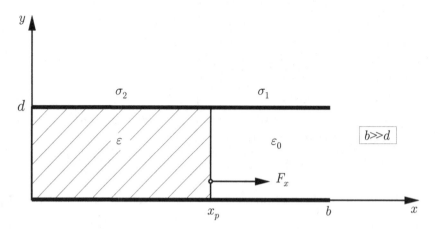

Before the dielectric has been inserted the charge density on the plates is $\pm\sigma_0 = \pm D_0 = \pm\varepsilon_0 E_0$ and the total charge is $\pm Q_0 = \pm\sigma_0 b^2$. The energy stored in the electric field $W_0$ is given by the integral of the energy density.

$$W_0 = \frac{1}{2} E_0 D_0 \cdot b^2 d = \frac{1}{2}\varepsilon_0 \frac{U_0^2}{d^2} b^2 d = \frac{1}{2} U_0^2 C_0 ; \qquad C_0 = \varepsilon_0 \frac{b^2}{d}$$

After inserting the dielectric the electric flux density $D$ is different in rooms of different materials and therefore the associated charge density $\sigma$ is also different. This is because the electric field is continuous at the boundary $x = x_p$, thus the electric field is identical everywhere between the plates. Its value depends on the position of the dielectric and therewith on the value $x_p$. As the total charge on the plates remains the same, we get the following result with two different charge densities $\sigma_1$ and $\sigma_2$ for the domains with permittivities $\varepsilon_0$ and $\varepsilon$.

$$\sigma_1(b - x_p) + \sigma_2 x_p = \sigma_0 b ; \qquad E_1 = E_2 = E$$

$$\varepsilon_0 \varepsilon E_1 = \varepsilon_0 \varepsilon E_2 = \varepsilon D_1 = \varepsilon \sigma_1 = \varepsilon_0 D_2 = \varepsilon_0 \sigma_2$$

$$\sigma_2 = \frac{\varepsilon}{\varepsilon_0}\sigma_1 ; \qquad \sigma_1 = \frac{\sigma_0}{1 + \dfrac{x_p}{b}\left(\dfrac{\varepsilon}{\varepsilon_0} - 1\right)} = \frac{\varepsilon_0}{\varepsilon}\sigma_2 = D_1$$

$$D_1 = \varepsilon_0 E = \frac{\varepsilon_0 E_0}{1 + \dfrac{x_p}{b}\left(\dfrac{\varepsilon}{\varepsilon_0} - 1\right)} = \frac{\varepsilon_0}{\varepsilon} D_2$$

$$E = \frac{E_0}{1 + \dfrac{x_p}{b}\left(\dfrac{\varepsilon}{\varepsilon_0} - 1\right)}$$

The limits $x_p = 0, b$ and $\varepsilon = \varepsilon_0$ provide the correct results.

The stored energy is

$$W(x_p) = \frac{1}{2}\varepsilon_0 E^2 db(b - x_p) + \frac{1}{2}\varepsilon E^2 dbx_p$$

$$= \frac{1}{2} E^2 db\left[\varepsilon_0(b - x_p) + \varepsilon x_p\right] = \frac{1}{2}\varepsilon_0 E^2 db^2\left[1 + \frac{x_p}{b}\left(\frac{\varepsilon}{\varepsilon_0} - 1\right)\right]$$

$$= \frac{W_0}{1 + \dfrac{x_p}{b}\left(\dfrac{\varepsilon}{\varepsilon_0} - 1\right)} .$$

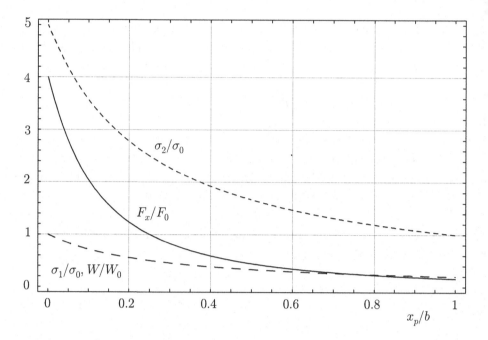

Fig. 2.14–1: Charges $\sigma_{1,2}/\sigma_0$ and force $F_x/F_0$ in dependence on the filling ratio $x_p/b$
with $\varepsilon/\varepsilon_0 = 5$; $F_0 = W_0/b$

With this result and the relation $\vec{F} = -\operatorname{grad} W$ the solution for the $x$–directed force
$F_x$ acting on the dielectric becomes

$$F_x = -\frac{\partial W}{\partial x_p} = \frac{W_0/b\left(\dfrac{\varepsilon}{\varepsilon_0} - 1\right)}{\left[1 + \dfrac{x_p}{b}\left(\dfrac{\varepsilon}{\varepsilon_0} - 1\right)\right]^2} > 0; \quad F_0 = W_0/b.$$

The dielectric is pulled into the capacitor.

## 2.15 2D-Problem with Homogeneous Boundary Conditions on Different Cartesian Coordinates

The drawing below shows the problem with a given potential $V_0(x)$ on the boundary
$y = 0$ in dependence on the $x$-coordinate, whereas all other boundaries own the potential $V = 0$.

Calculate the potential inside the closed boundary.

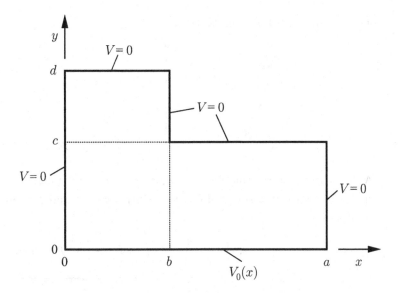

It is useful to split the computation area by the plane $y = c$, because of the given potential in dependence on the $x$-coordinate and the homogeneous boundary values $V = 0$ on $x = (0; a)$ and $x = (0; b)$. The potentials $V_{1,2}$ in the domains $0 \leq y \leq c$ and $c \leq y \leq d$ satisfy Laplace's equation $\Delta V_{1,2} = 0$. The functions which satisfy the homogeneous boundary conditions are trigonometric functions $\sin(p_{1,2,n}x)$ with $p_{1n} = n\pi/a$ and $p_{2n} = n\pi/b$. For the $y$-coordinate the hyperbolic functions are chosen, so that in the area $c \leq y \leq d$ the potential $V_2(x,y)$ at $y = d$ vanishes and in the area $0 \leq y \leq c$ the potential $V_1(x,y)$ is a combination of two linear independent parts, where the first has zeros at $y = 0$ and the second at $y = c$ for $b < x < a$.

$$V_1(x,y) = \sum_{n=1}^{\infty} \left[ A_n \frac{\sinh(p_{1n}y)}{\sinh(p_{1n}c)} + B_n \frac{\sinh(p_{1n}(y-c))}{\sinh(p_{1n}(-c))} \right] \sin(p_{1n}x) \; ; \qquad 0 \leq y \leq c$$

$$V_2(x,y) = \sum_{n=1}^{\infty} C_n \sin(p_{2n}x) \frac{\sinh(p_{2n}(y-d))}{\sinh(p_{2n}(c-d))} \; ; \qquad c \leq y \leq d$$

On the surface $y = 0$ applies

$$V_1(x, y = 0) = V_0(x) = \sum_{n=1}^{\infty} B_n \sin(p_{1n}x)$$

$$B_n = \frac{2}{a} \int_0^a V_0(x) \sin(p_{1n}x) \, dx \, .$$

The continuity of the potential in $y = c$ leads to

$$V_1(x, y = c) = \begin{cases} 0 & ; \quad b \le x \le a \\ V_2(x, y = c) & ; \quad 0 \le x \le b \end{cases}$$

$$\sum_{n=1}^{\infty} A_n \sin(p_{1n}x) = \begin{cases} 0 & ; \quad b \le x \le a \\ \sum_{n=1}^{\infty} C_n \sin(p_{2n}x) & ; \quad 0 \le x \le b \end{cases}.$$

As the trigonometric functions $\sin(p_{1n}x)$ form an orthogonal system of functions, multiplication with an associated function $\sin(p_{1m}x)$ and integration over $x = 0$ to $x = a$ results in

$$\frac{a}{2} A_m = \sum_{n=1}^{\infty} C_n \underbrace{\int_0^b \sin(p_{2n}x) \sin(p_{1m}x) dx}_{\alpha_{nm}}$$

and therewith we get the following system of equations

$$A_m = \frac{2}{a} \sum_{n=1}^{\infty} \alpha_{nm} C_n \; ; \qquad m = 1, 2, \dots .$$

The continuity of the normal component of the flux density in $y = c$ requires

$$\left. \frac{\partial V_1}{\partial y} \right|_{y=c} = \left. \frac{\partial V_2}{\partial y} \right|_{y=c} \; ; \qquad 0 \le x < b$$

$$\Rightarrow \sum_{n=1}^{\infty} \left[ A_n \coth(p_{1n}c) - \frac{B_n}{\sinh(p_{1n}c)} \right] p_{1n} \sin(p_{1n}x) =$$

$$= \sum_{n=1}^{\infty} C_n p_{2n} \sin(p_{2n}x) \coth(p_{2n}(c - d)) .$$

The trigonometric functions $\sin(p_{2n}x)$ also form an orthogonal system of functions, thus multiplication with an associated function $\sin(p_{2m}x)$ and integration over $x = 0$ bis $x = b$ results in

$$\frac{b}{2} C_m p_{2m} \coth(p_{2m}(c - d)) =$$

$$= \sum_{n=1}^{\infty} \left[ A_n \coth(p_{1n}c) - \frac{B_n}{\sinh(p_{1n}c)} \right] p_{1n} \cdot \underbrace{\int_0^b \sin(p_{1n}x) \sin(p_{2m}x) dx}_{\alpha_{mn}}$$

$$C_m = \frac{2}{m\pi \coth(p_{2m}(c-d))} \sum_{n=1}^{\infty} p_{1n}\alpha_{mn} \left[ A_n \coth(p_{1n}c) - \frac{B_n}{\sinh(p_{1n}c)} \right].$$

The potential and the field are determined, when both systems of equations are solved.

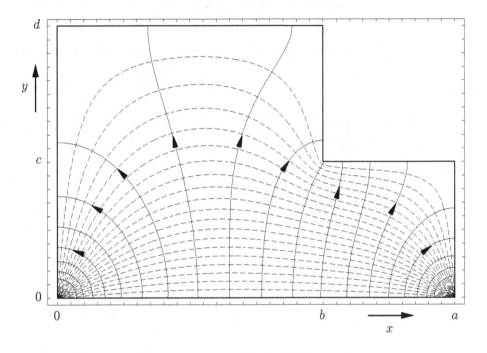

Fig. 2.15–1: Equipotential lines and electric lines of force for $b/a = 2/3$, $c/a = 1/3$, $d/a = 2/3$, and $V_0 = 1\,[\mathrm{V}]$

## 2.16 Method of Images for Dielectric Half-Spaces

The plane $z = 0$ divides the space into two half-spaces $z > 0$ of permittivity $\varepsilon_1$ and $z < 0$ of permittivity $\varepsilon_2$. The point $\vec{r}_{q1}$ in $z > 0$ marks the position of a point charge $Q_1$ and the point $\vec{r}_{q2} = \vec{r}_{q1} - 2\vec{e}_z(\vec{r}_{q1}\vec{e}_z) = \vec{r}_{q1}^*$ in $z < 0$ marks a second point charge $Q_2$.

Calculate the force acting on the charges.

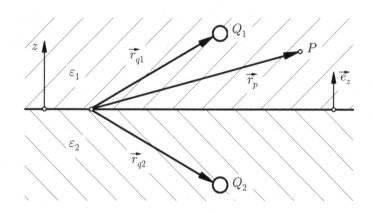

If only the point charge $Q_1$ in $z > 0$ is considered, then the method of images for dielectric half-spaces yields the potential

$$V(\vec{r}_p) = \frac{Q_1}{4\pi\varepsilon_1} \begin{cases} \dfrac{1}{|\vec{r}_p - \vec{r}_{q1}|} - \dfrac{k}{|\vec{r}_p - \vec{r}^*_{q1}|}; & z \geq 0; \quad k = (\varepsilon_2 - \varepsilon_1)/(\varepsilon_2 + \varepsilon_1) \\[4mm] \dfrac{1-k}{|\vec{r}_p - \vec{r}_{q1}|}; & z \leq 0; \quad \vec{r}^*_{q1,2} = \vec{r}_{q1,2} - 2\vec{e}_z(\vec{e}_z\vec{r}_{q1,2}). \end{cases}$$

Accordingly in the presence of the second point charge $Q_2$ in the half-space $z < 0$ the total solution for the potential is

$$V(\vec{r}_p) = \frac{Q_1}{4\pi\varepsilon_1} \left[ \frac{1}{|\vec{r}_p - \vec{r}_{q1}|} - \frac{k}{|\vec{r}_p - \vec{r}^*_{q1}|} \right] + \frac{Q_2}{4\pi\varepsilon_2} \frac{1+k}{|\vec{r}_p - \vec{r}_{q2}|}; \quad z \geq 0$$

$$V(\vec{r}_p) = \frac{Q_1}{4\pi\varepsilon_1} \frac{1-k}{|\vec{r}_p - \vec{r}_{q1}|} + \frac{Q_2}{4\pi\varepsilon_2} \left[ \frac{1}{|\vec{r}_p - \vec{r}_{q2}|} + \frac{k}{|\vec{r}_p - \vec{r}^*_{q2}|} \right]; \quad z \leq 0.$$

We can proof the result for the case $\varepsilon_1 = \varepsilon_2 = \varepsilon$, thus $k = 0$, and get the expected result

$$V(\vec{r}_p) = \frac{Q_1}{4\pi\varepsilon} \frac{1}{|\vec{r}_p - \vec{r}_{q1}|} + \frac{Q_2}{4\pi\varepsilon} \frac{1}{|\vec{r}_p - \vec{r}_{q2}|}.$$

With $\varepsilon_2 \gg \varepsilon_1$, thus $k \to 1$, we get

$$V(\vec{r}_p) = \frac{Q_1}{4\pi\varepsilon_1} \left[ \frac{1}{|\vec{r}_p - \vec{r}_{q1}|} - \frac{1}{|\vec{r}_p - \vec{r}^*_{q1}|} \right]; \quad z > 0; \quad V = 0; \quad z < 0.$$

This result agrees with the method of images for the conducting half-space. In the same way with $\varepsilon_1 \gg \varepsilon_2$ and $k \to -1$ we get

$$V = 0; \quad z > 0; \quad V(\vec{r}_p) = \frac{Q_2}{4\pi\varepsilon_2} \left[ \frac{1}{|\vec{r}_p - \vec{r}_{q2}|} - \frac{1}{|\vec{r}_p - \vec{r}^*_{q2}|} \right]; \quad z < 0.$$

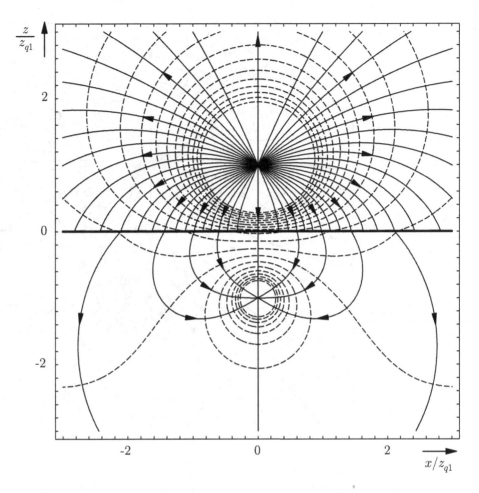

Fig. 2.16–1: Equipotential lines and electric lines of force for $Q_1/Q_2 = -2$,
$\varepsilon_1/\varepsilon_2 = 1/3$, and $z_{q2} = -z_{q1}$

The forces $\vec{F}_{1,2} = \vec{e}_z \, F_{1,2}$ acting on the charges $Q_{1,2}$ are

$$F_1 \;=\; \frac{Q_1}{4\pi\varepsilon_1}\,\frac{-kQ_1 + \dfrac{\varepsilon_1}{\varepsilon_2}(1+k)Q_2}{(z_1 - z_2)^2}\;; \qquad F_2 \;=\; \frac{Q_2}{4\pi\varepsilon_2}\,\frac{-kQ_2 - \dfrac{\varepsilon_2}{\varepsilon_1}(1-k)Q_1}{(z_1 - z_2)^2}\,.$$

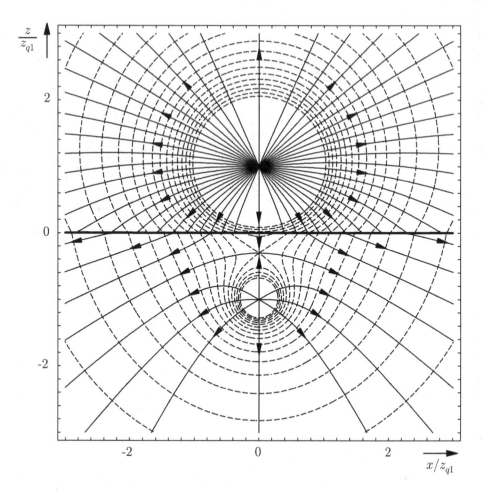

Fig. 2.16–2: Equipotential lines and electric lines of force for $Q_1/Q_2 = 2$,
$\varepsilon_1/\varepsilon_2 = 1/3$, and $z_{q2} = -z_{q1}$

## 2.17 Concentric Cylinders with Given Potentials

The cylinder of radius $\varrho = b$ has the potential $V_1$. Within this cylinder a second
cylinder of radius $\varrho = a < b$ has the potentials

$$V(\varrho = a, 0 < \varphi < \pi) = V_0 \quad \text{and} \quad V(\varrho = a, \pi < \varphi < 2\pi) = -V_0.$$

Find the electric field in the whole space. The permittivity $\varepsilon$ is constant.

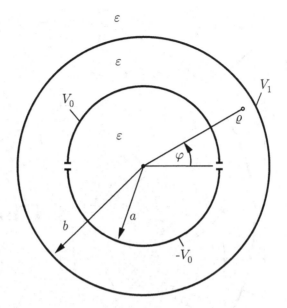

The solution of Laplace's equation $\Delta V(\varrho, \varphi) = 0$ is

$$
V(\varrho, \varphi) = \begin{cases}
\displaystyle\sum_{n=1}^{\infty} A_n (\varrho/a)^n \sin(n\varphi) & ; \quad \varrho \le a \\[2ex]
\begin{aligned} & B_0 + C_0 \ln(\varrho)/\ln(a) + \\ & + \sum_{n=1}^{\infty} [B_n(\varrho/a)^n + C_n(a/\varrho)^n] \sin(n\varphi) \end{aligned} & ; \quad a \le \varrho \le b \\[3ex]
V_1 \ln(\varrho)/\ln(b) & ; \quad \varrho \ge b.
\end{cases}
$$

Evaluating the boundary condition in $\varrho = a$ leads to

$$
\sum_{n=1}^{\infty} A_n \sin(n\varphi) = B_0 + C_0 + \sum_{n=1}^{\infty} [B_n + C_n] \sin(n\varphi) = \begin{cases} V_0 ; & 0 < \varphi < \pi \\ -V_0 ; & \pi < \varphi < 2\pi. \end{cases}
$$

With the orthogonality relation for trigonometric functions we derive by multiplication with $\sin(m\varphi)$ and integration from $\varphi = 0$ to $\varphi = 2\pi$ the equations

$$
B_0 + C_0 = 0 ;
$$

$$
A_n \pi = (B_n + C_n)\pi = 2V_0 \int_0^{\pi} \sin(n\varphi) d\varphi =
$$

$$
= \frac{2V_0}{n}[1 - \cos(n\pi)] = \frac{4V_0}{n} ; \qquad n = 2k+1 ; \quad k = 0, 1, 2, \dots
$$

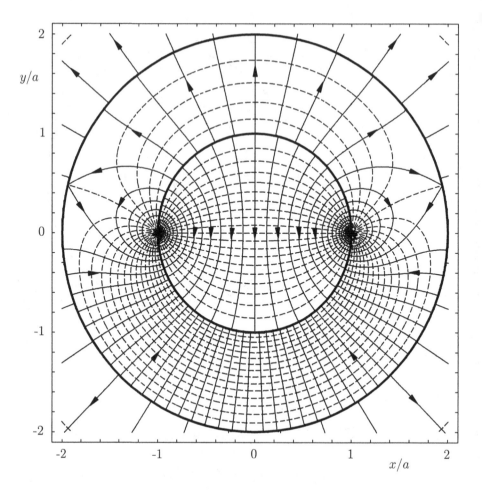

Fig. 2.17–1: Equipotential lines and electric lines of force for $b/a = 2$ and $V_1/V_0 = 0.5$

$$\Rightarrow \quad A_n = \frac{4V_0}{n\pi} = B_n + C_n ; \qquad n = 2k+1 .$$

On the cylinder $\varrho = b$ applies

$$B_0 + C_0 \frac{\ln(b)}{\ln(a)} + \sum_{n=1}^{\infty} [B_n(b/a)^n + C_n(a/b)^n] \sin(n\varphi) = V_1$$

and hence

$$B_0 + C_0 \ln(b)/\ln(a) = V_1 ; \qquad B_n(b/a)^n + C_n(a/b)^n = 0 .$$

Finally the constants are

$$B_0 = \frac{V_1}{1 - \ln(b)/\ln(a)} ; \qquad C_0 = -B_0$$

$$B_n = \frac{4V_0}{n\pi[1-(b/a)^{2n}]} \;; \qquad C_n = -B_n(b/a)^{2n} \;.$$

The electric field $\vec{E}$ is

$$\vec{E} = -\operatorname{grad} V = -\vec{e}_\varrho \frac{\partial V}{\partial \varrho} - \frac{\vec{e}_\varphi}{\varrho}\frac{\partial V}{\partial \varphi} = \vec{e}_\varrho E_\varrho + \vec{e}_\varphi E_\varphi \;.$$

$$E_\varrho = -\begin{cases} \sum\limits_{k=0}^{\infty} A_n\, n/\varrho\,(\varrho/a)^n \sin(n\varphi) & ; \varrho \le a \\[2mm] \frac{C_0}{\varrho \ln(a)} + \sum\limits_{k=0}^{\infty} n/\varrho[B_n(\varrho/a)^n - C_n(a/\varrho)^n]\sin(n\varphi) & ; a \le \varrho \le b \\[2mm] V_1/(\varrho \ln(b)) & ; \varrho \ge b \end{cases}$$

$$E_\varphi = \begin{cases} -\sum\limits_{k=0}^{\infty} n/\varrho \begin{cases} A_n(\varrho/a)^n \\[1mm] B_n(\varrho/a)^n + C_n(a/\varrho)^n \end{cases}\cos(n\varphi) & \begin{matrix} ; \varrho \le a \\[1mm] ; a \le \varrho \le b, \end{matrix} \\[3mm] 0; & ; \varrho \ge b \end{cases}$$

$$\text{with} \quad n = 2k+1.$$

## 2.18   Force on a Ring Charge inside a Conducting Cylinder

A ring charge $\lambda_q$ of radius $b$ is concentrically located on the plane $z = c > 0$ inside a conducting and grounded cylinder of infinite length and radius $a$. The domain $z < 0$ has the permittivity $\varepsilon$, whereas the domain $z > 0$ has the permittivity $\varepsilon_0$.

What is the force that acts on the ring charge?

In the limit $\varepsilon = \varepsilon_0$ no force acts on the ring charge. For the calculation of the force in case of $\varepsilon \ne \varepsilon_0$ we need knowledge of the variation in the electric field, when the permittivity in the area $z < 0$ changes from $\varepsilon_0$ to $\varepsilon$. For this purpose we first calculate the exciting potential of the ring charge within homogeneous permittivity.

The exciting potential $V_E$ of the ring charge is calculated under the assumption $\varepsilon = \varepsilon_0$. With cylindrical coordinates the solution of Laplace's equation $\Delta V_E(\varrho, z) = 0$, $z \ne c$, requires Bessel functions of order zero.

$$V_E(\varrho, z) = \sum_{r=1}^{\infty} A_r\, J_0(x_{0r}\,\varrho/a)\, \exp(-x_{0r}|z - c|/a)$$

The ansatz with the zeros $x_{0r}$ of the Bessel functions satisfies the boundary condition $V_E(\varrho = a, z) = 0$ and the exponential function describes the decay of the potential away from the charge. It is useful to express the ring charge in terms of a surface charge

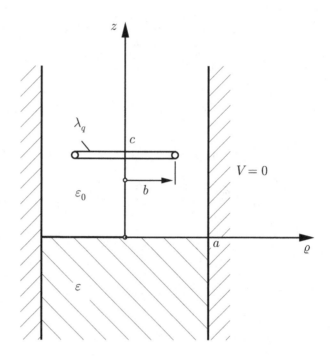

$\sigma(\varrho)$ in the plane $z = c$, where the charge density is zero except for points on the ring $(\varrho = b; z = c)$. With the charge $\sigma(\varrho)$ the boundary condition for the $z$-component of the electric flux density $\vec{D}$ in $z = c$ leads to

$$\vec{e}_z \left( \vec{D} \bigg|_{\substack{z>c \\ z \to c}} - \vec{D} \bigg|_{\substack{z<c \\ z \to c}} \right) = \sigma(\varrho)$$

$$\Leftrightarrow \quad \frac{\partial V}{\partial z} \bigg|_{\substack{z<c \\ z \to c}} - \frac{\partial V}{\partial z} \bigg|_{\substack{z>c \\ z \to c}} = \frac{\sigma(\varrho)}{\varepsilon_0}$$

$$\Rightarrow \quad \frac{2}{a} \sum_{r=1}^{\infty} x_{0r} A_r J_0(x_{0r}\varrho/a) \doteq \frac{1}{\varepsilon_0} \sigma(\varrho).$$

The Bessel functions $J_0(x_{0r}\varrho/a)$ belong to an orthogonal system of functions, thus the multiplication with an orthogonal function $J_0(x_{0s}\varrho/a)$ and the weight function $\varrho$ results after integration from $\varrho = 0$ to $\varrho = a$ in

$$\frac{2}{a} x_{0s} A_s \frac{1}{2} a^2 J_1^2(x_{0s}) = \frac{1}{\varepsilon_0} \int_0^a \sigma(\varrho) J_0(x_{0s}\varrho/a) \varrho \, d\varrho,$$

where the orthogonality relation for Bessel functions has been applied. For the remaining integral the integration interval restricts to the location of the ring charge $\varrho = b \pm \delta$

and we get

$$
A_r = \frac{1}{\varepsilon_0 x_{0r} a J_1^2(x_{0r})} \int\limits_{b-\delta}^{b+\delta} \sigma(\varrho) \, J_0(x_{0r}\varrho/a) \, \varrho d\varrho
$$

$$
= \frac{J_0(x_{0r}b/a)b}{\varepsilon_0 a x_{0r} J_1^2(x_{0r})} \underbrace{\int\limits_{b-\delta}^{b+\delta} \sigma(\varrho) d\varrho}_{\lambda_q} \qquad \text{with} \quad \delta \to 0
$$

$$
A_r = V_0 \frac{J_0(x_{0r}b/a)}{x_{0r} J_1^2(x_{0r})} \; ; \qquad V_0 = \frac{2\pi\lambda_q b}{2\pi\varepsilon_0 a} = \frac{Q}{2\pi\varepsilon_0 a} = E_0 a \, .
$$

The result for the exciting potential is

$$
V_E(\varrho, z) = V_0 \sum_{r=1}^{\infty} \frac{J_0(x_{0r}b/a)}{x_{0r} J_1^2(x_{0r})} J_0(x_{0r}\varrho/a) \exp(-x_{0r} |z-c| /a) \, .
$$

When the space $z < 0$ has the permittivity $\varepsilon$, then the variation in the field results from the difference between the exciting potential and the resulting potential $V(\varrho, z)$.

$$
V(\varrho, z) = V_E(\varrho, z) + V_0 \sum_{r=1}^{\infty} B_r \frac{J_0(x_{0r}b/a)}{x_{0r} J_1^2(x_{0r})} J_0(x_{0r}\varrho/a) \exp(-x_{0r}|z|/a)
$$

The additional potential in this approach uses the solution for the exciting potential, but applies different constants $B_r$. The total approach already satisfies all boundary conditions for the potential and also for the electric flux density in $z = c$. Finally evaluating the continuity relation of the flux density in the boundary layer of different permittivities $z = 0$ leads to the constants $B_r$.

$$
\varepsilon_0 \left. \frac{\partial V}{\partial z} \right|_{\substack{z>0 \\ z\to 0}} = \varepsilon \left. \frac{\partial V}{\partial z} \right|_{\substack{z<0 \\ z\to 0}}
$$

$$
\Leftrightarrow \quad \varepsilon_0 \left[ \exp(-x_{0r}c/a) - B_r \right] = \varepsilon \left[ \exp(-x_{0r}c/a) + B_r \right]
$$

$$
\Rightarrow \quad B_r = -k \, \exp(-x_{0r}c/a) \; ; \qquad k = \frac{\varepsilon - \varepsilon_0}{\varepsilon + \varepsilon_0}
$$

$$
V(\varrho, z) = V_E(\varrho, z) - kV_0 \sum_{r=1}^{\infty} \frac{J_0(x_{0r}b/a)}{x_{0r} J_1^2(x_{0r})} J_0(x_{0r}\varrho/a) \cdot
$$

$$
\cdot \exp(-x_{0r}c/a) \begin{cases} \exp(-x_{0r}z/a) & ; \quad z \geq 0 \\ \exp(x_{0r}z/a) & ; \quad z \leq 0 \end{cases}
$$

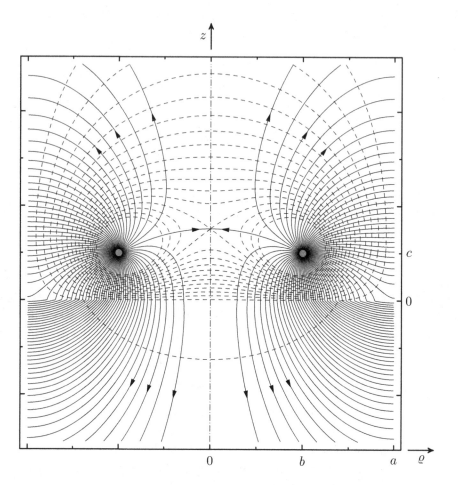

Fig. 2.18–1: Equipotential lines and electric lines of force for $k = 0.9$ and $a = 2b = 4c$

$$\Rightarrow \quad V(\varrho, z) = \begin{cases} V_E(\varrho, z) - kV_E(\varrho, -z) & ; \quad z \geq 0 \\ (1-k)\, V_E(\varrho, z) & ; \quad z \leq 0 \end{cases}$$

This result agrees with the method of images for dielectric half spaces.

The force $\vec{F}$ on the ring charge acts in axial direction.

$$\vec{F} = \vec{e}_z F \, ; \qquad F = 2\pi b\, \lambda_q E_z\big|_{\substack{z=c \\ \varrho=b}} = -2\pi b\, \lambda_q \left.\frac{\partial V_s}{\partial z}\right|_{\substack{z=c \\ \varrho=b}}$$

$$\text{with} \quad V_s = V(\varrho, z) - V_E(\varrho, z)$$

$$\Rightarrow \quad F = -2\pi b \lambda_q k V_0/a \sum_{r=1}^{\infty} \frac{J_0^2(x_{0r}b/a)}{J_1^2(x_{0r})} \exp(-2x_{0r}c/a)$$

$$= -k Q E_0 \sum_{r=1}^{\infty} \frac{J_0^2(x_{0r}b/a)}{J_1^2(x_{0r})} \exp(-2x_{0r}c/a)$$

With a constant total charge $Q$ the limit $b \to 0$ results in

$$F\bigg|_{\substack{b \to 0 \\ Q = \text{const}}} = -k Q E_0 \sum_{r=1}^{\infty} \frac{\exp(-2x_{0r}c/a)}{J_1^2(x_{0r})}.$$

If in addition the limit case $c \to 0$ is considered, we get an divergent series and thus an unbounded force. Within a boundary layer of different permittivities only surface charges can exist.

## 2.19 Geometry with Circular Symmetry and Given Potentials on Parallel Planes

The potential $V(\varrho, z = \pm a) = \pm V_0$ is given. On the cylindrical surface $\varrho = c$ between the planes the potential is $V(\varrho = c, z) = 0$ for $|z| < a$.

Find the potential $V(\varrho, z)$ in the domain $(\varrho \geq c;\ |z| \leq a)$.

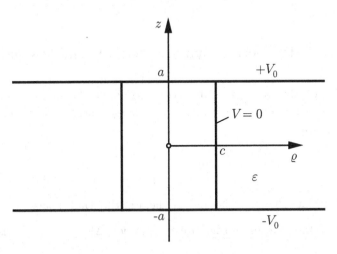

We consider a potential $V_s(\varrho, z) = V(\varrho, z) - V_0 z/a$, which has to satisfy the boundary conditions $V_s(\varrho, z = \pm a) = 0$ and $V_s(\varrho = c, z) = -V_0 z/a$.

Because the potential on $\varrho = c$ depends on the $z$-coordinate, the solution of the differential equation $\Delta V_s = 0$ is

$$V_s(\varrho, z) = \sum_{n=1}^{\infty} A_n \frac{I_0(p_n\varrho)}{I_0(p_nc)} \sin(p_nz).$$

The boundary condition for the potential $V_s$ at points $z = \pm a$ requires

$$\sin(p_na) = 0 ; \qquad p_n = n\pi/a ; \quad n = 1,2,3,\ldots$$

and at $\varrho = c$

$$-V_0\, z/a = \sum_{n=1}^{\infty} A_n \sin(p_nz).$$

With the orthogonality relation for trigonometric functions we get

$$A_n\, a = -V_0 \int_{-a}^{a} \frac{z}{a} \sin(p_nz)\, dz ; \qquad A_n = -\frac{V_0}{(p_na)^2} \int_{-p_na}^{p_na} x \sin(x)\, dx$$

$$A_n = \frac{2V_0}{p_na} (-1)^n$$

and finally the resulting potential is

$$V(\varrho, z) = V_0 \left[ \frac{z}{a} + 2 \sum_{n=1}^{\infty} \frac{(-1)^n}{p_na} \frac{I_0(p_n\varrho)}{I_0(p_nc)} \sin(p_nz) \right].$$

## 2.20   Dielectric Cylinder with Variable Charge on its Surface

Within the space of permittivity $\varepsilon_0$ the cylinder $\varrho \le a$ is filled with material of permittivity $\varepsilon$. On its surface $\varrho = a$ a charge distribution depending on the $\varphi$-coordinate is given.

$$\sigma(\varphi) = \begin{cases} \sigma_0 & |\varphi| < \varphi_0 \\ -\sigma_0 & \varphi_0 < \varphi < 2\pi - \varphi_0 \end{cases}$$

Calculate the potential and the field in the whole space.

Here the solution of the Laplace equation $\Delta V = 0$ in $\varrho \ne a$ is

$$V(\varrho, \varphi) = C_0 \left\{ \begin{array}{c} 1 \\ \ln\varrho/\ln a \end{array} \right\} + \sum_{n=1}^{\infty} C_n \left\{ \begin{array}{c} (\varrho/a)^n \\ (a/\varrho)^n \end{array} \right\} \cos(n\varphi) ; \qquad \begin{array}{c} \varrho \le a \\ \varrho \ge a \end{array} .$$

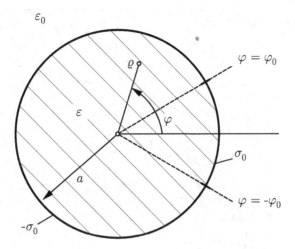

This approach already satisfies the continuity relation of the potential on the cylindrical surface $\varrho = a$.

The remaining constants follow from the boundary condition for the normal component $D_\varrho$ of the electric flux density $\vec{D}$ on the cylindrical surface.

$$D_\varrho\big|_{\substack{\varrho>a \\ \varrho\to a}} - D_\varrho\big|_{\substack{\varrho<a \\ \varrho\to a}} = \sigma(\varphi)$$

$$\Leftrightarrow \quad \varepsilon\,\frac{\partial V}{\partial \varrho}\bigg|_{\substack{\varrho<a \\ \varrho\to a}} - \varepsilon_0\,\frac{\partial V}{\partial \varrho}\bigg|_{\substack{\varrho>a \\ \varrho\to a}} = \sigma(\varphi)$$

$$\Rightarrow \quad -\varepsilon_0\,\frac{C_0}{a\ln a} + \sum_{n=1}^{\infty}\frac{nC_n}{a}(\varepsilon+\varepsilon_0)\cos(n\varphi) = \sigma(\varphi)$$

Integration over the coordinate $\varphi$ from $\varphi = 0$ to $\varphi = 2\pi$ leads do

$$C_0 = -\frac{a\ln a}{2\pi\varepsilon_0}\int_0^{2\pi}\sigma(\varphi)\,d\varphi\,; \qquad C_0 = \frac{\sigma_0 a\ln a}{\varepsilon_0}\left[1 - 2\frac{\varphi_0}{\pi}\right].$$

The constants $C_n, n \geq 1$ are calculated by means of the orthogonality relation for trigonometric functions.

$$\frac{nC_n}{a}\pi(\varepsilon+\varepsilon_0) = \int_0^{2\pi}\sigma(\varphi)\cos(n\varphi)\,d\varphi = \frac{\sigma_0}{n}\left[\sin(n\varphi)\big|_{-\varphi_0}^{\varphi_0} - \sin(n\varphi)\big|_{\varphi_0}^{2\pi-\varphi_0}\right]$$

$$\Rightarrow \quad C_n = \frac{4\sigma_0 a}{\pi(\varepsilon+\varepsilon_0)}\frac{\sin(n\varphi_0)}{n^2}$$

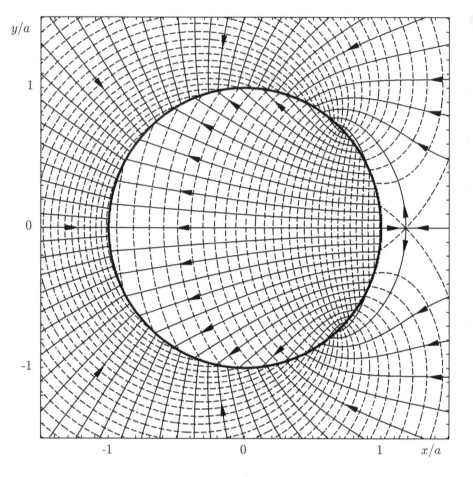

Fig. 2.20–1: Electric lines of force and equipotential lines for $\varphi_0 = \pi/4$ and $\varepsilon/\varepsilon_0 = 3/2$

The electric field $\vec{E} = -\operatorname{grad} V = -\vec{e}_\varrho \dfrac{\partial V}{\partial \varrho} - \dfrac{\vec{e}_\varphi}{\varrho} \dfrac{\partial V}{\partial \varphi} = \vec{e}_\varrho E_\varrho + \vec{e}_\varphi E_\varphi$ is

$$E_\varrho = -\left\{ \begin{array}{c} 0 \\ C_0/(\varrho \ln a) \end{array} \right\} - \sum_{n=1}^{\infty} \frac{n}{a} C_n \left\{ \begin{array}{c} (\varrho/a)^{n-1} \\ -(a/\varrho)^{n+1} \end{array} \right\} \cos(n\varphi) \; ; \qquad \begin{array}{c} \varrho < a \\ \varrho > a \end{array}$$

$$E_\varphi = \sum_{n=1}^{\infty} \frac{n}{a} C_n \left\{ \begin{array}{c} (\varrho/a)^{n-1} \\ (a/\varrho)^{n+1} \end{array} \right\} \sin(n\varphi) \; ; \qquad \begin{array}{c} \varrho < a \\ \varrho > a \end{array} \; .$$

## 2.21 Potential and Field of Dipole Layers

The sphere $r = a$ carries an electric dipole layer of density $\vec{m} = \pm m\,\vec{r}/r$. In the area $0 \le \vartheta < \vartheta_0$ the sign of the radially directed dipoles is positive, whereas in the remaining area $\vartheta_0 < \vartheta \le \pi$ the sign is negative. The absolute value $m$ and the permittivity $\varepsilon$ are constant.

Calculate the potential and the field at points on the $z$-axis.

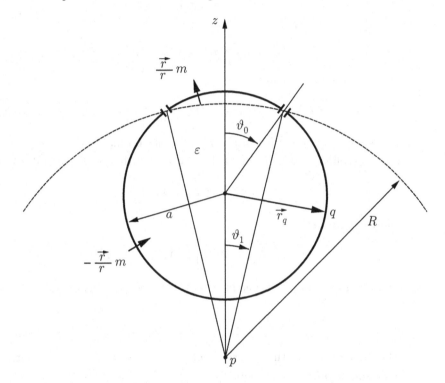

The potential of a homogeneous dipole moment density in normal direction on the surface $a_D$ is given by

$$V(\vec{r}_p) = \frac{m}{4\pi\varepsilon} \int\limits_{a_D} \frac{\vec{r}}{r^3}\, d\vec{a}_D = \frac{m}{4\pi\varepsilon} \int\limits_{a_D} \vec{n}\frac{\vec{r}}{r^3} da_D = -\frac{m}{4\pi\varepsilon}\Omega; \quad \vec{r} = \vec{r}_p - \vec{r}_q,$$

where $\Omega$ is the solid angle of the surface $a_D$ defined by

$$\Omega = -\int\limits_{a_D} \vec{n}\,\vec{r}/r^3 da_D \,.$$

It relates to the surface area on the sphere with radius 1 and origin $p$, which results from the projection of the boundary contour of the surface $a_D$ on the unit sphere.

This result shows, that the solid angle depends only on the boundary of the surface $a_D$ and on the orientation of its normal $\vec{n}$ to the observation point $p$. Therefore the potential of the dipole layer $m\,\vec{r}/r$ at $r = a$ and $0 \leq \vartheta < \vartheta_0$ on the $z$-axis in the interval $-\infty < z < a$ can also be calculated by means of the solid angle of a spherical calotte with radius $R$, the same boundary contour like $a_D$, and with an apex angle $2\vartheta_1$. In this way the solid angle in $z < a$ is

$$\Omega = -\int_{a_D} \vec{n}\,\vec{r}/r^3 da_D = 2\pi(1 - \cos\vartheta_1) = 2\pi\left[1 - \frac{-z + a\cos\vartheta_0}{R}\right]$$

$$R^2 = (-z + a\cos\vartheta_0)^2 + (a\sin\vartheta_0)^2 = z^2 + a^2 - 2az\cos\vartheta_0$$

and the potential becomes

$$V(z) = -\frac{m}{2\varepsilon}[1 + Z_0/R]\;; \quad z < a\;; \qquad Z_0 = z - a\cos\vartheta_0\,.$$

For points $z > a$ the orientation of the surface normal to the observation point changes, thus the solid angel of the dipole layer with coordinates $r = a$ and $0 \leq \vartheta < \vartheta_0$ becomes

$$\Omega = -2\pi\left[1 - Z_0/R\right]\;; \qquad z > a$$

and the potential is

$$V = \frac{m}{2\varepsilon}[1 - Z_0/R] = V_0\left[1 - Z_0/R\right]/2\;; \qquad z > a\,.$$

At the position of the dipole layer $z = a$ the potential changes by

$$V\big|_{\substack{z>a \\ z\to a}} - V\big|_{\substack{z<a \\ z\to a}} = \frac{m}{2\varepsilon} - \left(-\frac{m}{2\varepsilon}\right) = \frac{m}{\varepsilon}\,.$$

The dipole layer on the spherical calotte $r = a$ and $\vartheta_0 < \vartheta \leq \pi$ with density $-\vec{r}/r\,m$ analogously provides the potentials

$$V = -\frac{m}{2\varepsilon}[1 + Z_0/R] = -V_0\left[1 + Z_0/R\right]/2\;; \qquad z < -a$$

and

$$V = \frac{m}{2\varepsilon}[1 - Z_0/R] = V_0\left[1 - Z_0/R\right]/2\;; \qquad z > -a\,.$$

The superposition of the potentials gives the total potential

$$V(z) = -\frac{m}{\varepsilon}Z_0/R + \begin{cases} m/\varepsilon & ;\quad z > a \\ 0 & ;\quad -a < z < a \\ -m/\varepsilon & ;\quad z < -a\,. \end{cases}$$

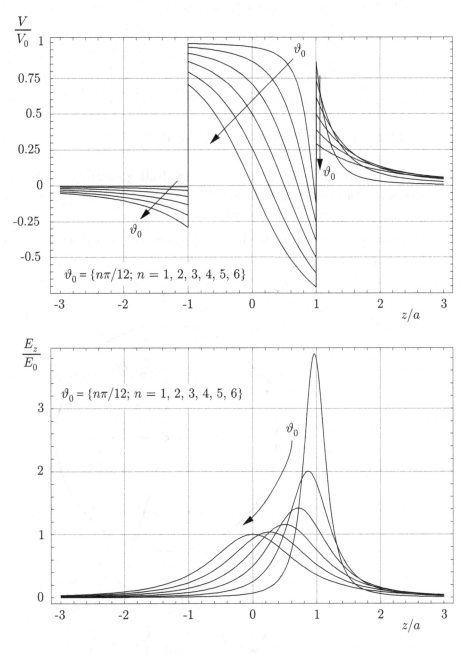

Fig. 2.21–1: Potential $V/V_0$ and field $E_z/E_0$ on the $z$-axis
$(V_0 = m/\varepsilon; \ E_0 = m/(a\varepsilon))$

Due to the rotational symmetry the electric field is $z$-directed and is derived by

$$\vec{E} = -\operatorname{grad} V; \quad E_z(z) = -\frac{\partial V}{\partial z} = \frac{m}{\varepsilon}\frac{\partial}{\partial z}\left(\frac{Z_0}{R}\right); \quad E_0 = \frac{V_0}{a}; \quad z \neq \pm a$$

$$E_z = \frac{m}{\varepsilon}\left[\frac{1}{R} - (z - a\cos\vartheta_0)^2\frac{1}{R^3}\right] = E_0\frac{a}{R}\left[1 - \left(\frac{Z_0}{R}\right)^2\right]; \quad z \neq \pm a.$$

## 2.22 Sphere with Given Potential

The area $0 \leq \vartheta < \vartheta_0$ of the spherical conducting surface $r = a$ has the potential $V_0$, whereas the remaining part $\vartheta_0 < \vartheta \leq \pi$, isolated from the first one, has the Potential $-V_0$. The permittivity $\varepsilon$ is constant.

Calculate the potential in the whole space and the charge density on the subarea $(r = a; 0 \leq \vartheta < \vartheta_0)$.

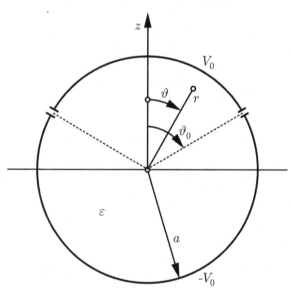

Because of the rotational symmetry with $\partial V/\partial\varphi = 0$ the solution functions of Laplace's equation $\Delta V(r, \vartheta, \varphi) = 0$ using spherical coordinates $(r, \vartheta, \varphi)$ are

$$V(r, \vartheta) = \sum_{n=0}^{\infty} A_n \left\{ \begin{array}{l} (r/a)^n \\ (a/r)^{n+1} \end{array} \right\} P_n(\cos\vartheta); \quad \begin{array}{l} r \leq a \\ r \geq a \end{array}.$$

Since the rotation axis is part of the computation domain the functions $Q_n(\cos\vartheta)$ can be dropped.

With the substitution $u = \cos\vartheta$, $u_0 = \cos\vartheta_0$, and the orthogonality relation for the Legendre polynomials $P_n(u)$ it follows

$$\sum_{n=0}^{\infty} A_n \int_{-1}^{1} P_n(u)P_k(u)du = V_0 \left[\int_{u_0}^{1} P_k(u)du - \int_{-1}^{u_0} P_k(u)du\right]$$

$$A_k \frac{2}{2k+1} = V_0 \begin{cases} -2u_0 & ;\quad k = 0 \\ \dfrac{P_{k+1} - P_{k-1}}{2k+1}\bigg|_{u_0}^{1} - \dfrac{P_{k+1} - P_{k-1}}{2k+1}\bigg|_{-1}^{u_0} & ;\quad k > 0 \end{cases}$$

$$A_0 = -u_0 V_0$$

$$A_n = V_0/2 \big[P_{n+1}(1) - P_{n-1}(1) + P_{n+1}(-1) - P_{n-1}(-1)$$
$$- P_{n+1}(u_0) + P_{n-1}(u_0) - P_{n+1}(u_0) + P_{n-1}(u_0)\big]; \quad n > 0$$

with $\quad P_n(+1) = 1; \quad P_n(-1) = (-1)^n$

$$\Rightarrow \quad A_n = V_0\left[P_{n-1}(u_0) - P_{n+1}(u_0)\right]; \quad n > 0.$$

Therewith the potential is well-defined.

The inner and outer charge densities on the surface ($r = a, u_0 < u < 1$) follow from the normal component of the electric flux density.

$$\sigma_i = -\vec{e}_r\vec{D}\bigg|_{\substack{\varrho<a\\\varrho\to a}} = \varepsilon\frac{\partial V}{\partial r}\bigg|_{\substack{\varrho<a\\\varrho\to a}} \quad ; \quad \sigma_a = \vec{e}_r\vec{D}\bigg|_{\substack{\varrho>a\\\varrho\to a}} = -\varepsilon\frac{\partial V}{\partial r}\bigg|_{\substack{\varrho>a\\\varrho\to a}}$$

$$\sigma_i = \frac{\varepsilon}{a}\sum_{n=0}^{\infty} A_n n P_n(u) = \frac{\varepsilon V_0}{a}\sum_{n=0}^{\infty} n\left[P_{n-1}(u_0) - P_{n+1}(u_0)\right]P_n(u)$$

$$\sigma_a = \frac{\varepsilon}{a}\sum_{n=0}^{\infty} A_n(n+1) P_n(u) = \frac{\varepsilon V_0}{a}\sum_{n=0}^{\infty}(n+1)\left[P_{n-1}(u_0) - P_{n+1}(u_0)\right]P_n(u)$$

The formal calculation of the total charge leads to a non-convergent series, because of the singularity of the electric field at the discontinuity of the potential.

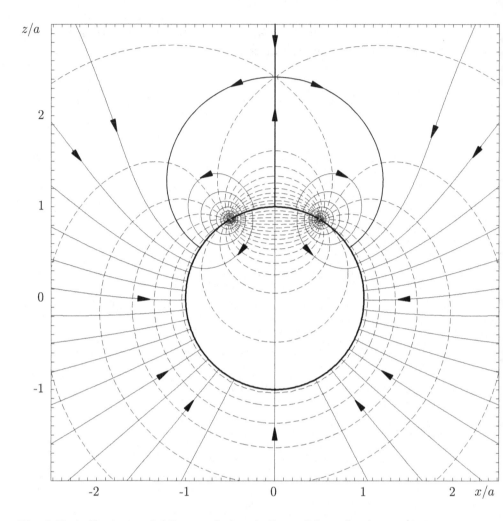

Fig. 2.22–1: Equipotential lines and electric lines of force for $\vartheta_0 = \pi/6$

## 2.23 Plane with Given Potential in Free Space

The given potential on the plane $y = 0$ depends only on the $x$-coordinate.

$$V(x, y = 0) = \left\{ \begin{array}{lll} V_0 & ; & |x| < a \\ 0 & ; & |x| > a \end{array} \right.$$

What is the potential in the whole space of permittivity $\varepsilon$?

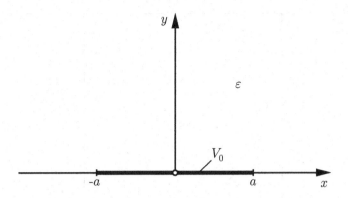

The potential satisfies the Laplace equation

$$\Delta V = \frac{\partial^2 V}{\partial x^2} + \frac{\partial^2 V}{\partial y^2} = 0 \; ; \qquad y \neq 0 \; ; \qquad \frac{\partial V}{\partial z} = 0 \, .$$

Because of the symmetry of the given potential $V(x,y) = V(-x,y)$ the following approach holds

$$V(x,y) = \int_0^\infty C(p) \, \cos(px) \, \exp(-p|y|) \, dp$$

with the spectral function $C(p)$.

The function of the given potential in $y = 0$ is expressible in terms of the Fourier integral

$$2V_0/\pi \int_0^\infty 1/p \, \sin(pa) \, \cos(px) \, dp = \begin{cases} V_0 & ; \quad |x| < a \\ 0 & ; \quad |x| > a \end{cases} .$$

Hence the spectral function becomes

$$C(p) = \frac{2V_0}{p\pi} \, \sin(pa)$$

and with the potential

$$V(x,y) = 2V_0/\pi \int_0^\infty 1/p \, \sin(pa) \, \cos(px) \, \exp(-p|y|) dp$$

the electric field is given by

$$\vec{E} = -\text{grad} \, V = 2V_0/\pi \int_0^\infty \sin(pa) \, [\vec{e}_x \, \sin(px) + \vec{e}_y \, \cos(px) \, \text{sign}\,(y)] \exp(-p|y|) \, dp \, .$$

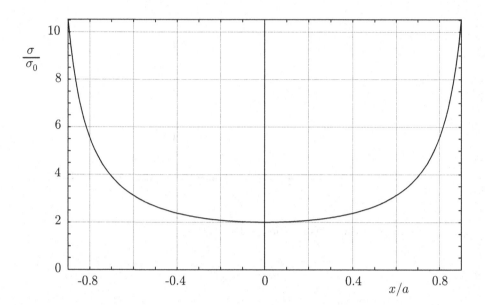

Fig. 2.23–1: Charge density for the interval $|x| < 0.9\,a$

In this special case the expressions for the integrals are known and it follows

$$V(x,y) \;=\; \frac{V_0}{\pi}\,[\arctan((a-x)/|y|) + \arctan((a+x)/|y|)]\;.$$

From this result the charge density $\sigma(x)$ can easily be deduced.

$$\sigma(x) \;=\; \varepsilon\left(\left.\frac{\partial V}{\partial y}\right|_{\substack{y<0\\y\to0\\x\neq\pm a}} - \left.\frac{\partial V}{\partial y}\right|_{\substack{y>0\\y\to0\\x\neq\pm a}}\right) \;=\; 2\varepsilon\frac{V_0}{\pi}\left[\frac{1}{a-x}+\frac{1}{a+x}\right]$$

$$\;=\; \frac{2\sigma_0}{1-(x/a)^2}\;;\quad \sigma_0 \;=\; \frac{2\varepsilon V_0}{\pi a}$$

The integral for the determination of the total charge per unit length $l$

$$Q/l \;=\; \int\limits_{-a}^{a} \sigma(x)dx$$

is unbounded. This is caused by the jump of the potential in $x=\pm a$.

## 2.24 Charge on a Plane between two Dielectrics

A cylindrical surface charge of density $\sigma = Q/(\pi a^2)$ is located in the plane $z = 0$ at points $\varrho \leq a$. This plane separates the half-spaces $z > 0$ with permittivity $\varepsilon_0$ and $z < 0$ with permittivity $\varepsilon$.

Calculate the potential and the field on the $z$-axis, which is an axis of symmetry. Furthermore analyze the limit case $a \to 0$ (point charge $Q$ in $(\varrho = 0 \, ; z = 0)$).

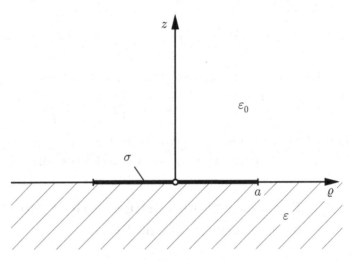

If the surface charge is located in a plane $z = c > 0$, then the calculation of the potential could be done by means of the method of images. In this case an additional image charge $-k\sigma$ with $k = (\varepsilon - \varepsilon_0)/(\varepsilon + \varepsilon_0)$ in the plane $z = -c$ must be considered to determine the potential in $z \geq 0$, whereas the permittivity is $\varepsilon_0$ in both half-spaces. The potential for points $z < 0$ results from a surface charge $(1 - k)\sigma$ in $z = c$. For the present problem the limit case $c \to 0$ holds, thus a surface charge $(1 - k)\sigma$ in $z = 0$ is the substitute in space of homogeneous permittivity $\varepsilon_0$.

Hence the potential on the $z$-axis is

$$V(z) = \frac{1}{4\pi\varepsilon_0} \int\limits_{a_S} \frac{(1-k)\sigma}{r} \, da_S = \frac{(1-k)Q}{4\pi^2\varepsilon_0 a^2} \int\limits_{\varphi=0}^{2\pi} \int\limits_{\varrho=0}^{a} \frac{\varrho \, d\varrho \, d\varphi}{\sqrt{z^2 + \varrho^2}} = \frac{(1-k)Q}{2\pi\varepsilon_0 a^2} \int\limits_{\zeta=z^2}^{z^2+a^2} \frac{d\zeta}{2\sqrt{\zeta}}$$

$$V(z) = \frac{(1-k)Q}{2\pi\varepsilon_0 a^2} \left[ \sqrt{z^2 + a^2} - |z| \right] \, .$$

The electric field results from

$$\vec{E} = -\operatorname{grad} V = \vec{e}_z \, E_z + \vec{e}_\varrho \, E_\varrho \, ; \qquad \vec{E}(\varrho = 0, z) = \vec{e}_z E_z = -\vec{e}_z \frac{\partial V}{\partial z}$$

and becomes

$$E_z = -\frac{\partial V}{\partial z} = -\frac{(1-k)Q}{2\pi\varepsilon_0 a^2} \left[ \frac{z}{\sqrt{z^2 + a^2}} + \left\{ \begin{array}{c} -1 \\ +1 \end{array} \right\} \right] ; \quad \begin{array}{c} z > 0 \\ z < 0 \end{array}$$

$$= -\frac{\sigma}{\varepsilon + \varepsilon_0} \left[ \frac{z}{\sqrt{z^2 + a^2}} + \left\{ \begin{array}{c} -1 \\ +1 \end{array} \right\} \right] ; \quad \begin{array}{c} z > 0 \\ z < 0 \end{array} .$$

The electric flux density $\vec{D}$ satisfies the condition

$$\vec{e}_z \left( \vec{D} \Big|_{\substack{z>0 \\ z\to 0}} - \vec{D} \Big|_{\substack{z<0 \\ z\to 0}} \right) = \sigma = \varepsilon \left. \frac{\partial V}{\partial z} \right|_{\substack{z<0 \\ z\to 0}} - \varepsilon_0 \left. \frac{\partial V}{\partial z} \right|_{\substack{z>0 \\ z\to 0}}$$

at points $\varrho < a$ on the surface charge in $z = 0$. It follows

$$\frac{(1-k)Q}{2\pi\varepsilon_0 a^2} (\varepsilon + \varepsilon_0) = \frac{Q}{2\pi\varepsilon_0 a^2} \left( 1 - \frac{\varepsilon - \varepsilon_0}{\varepsilon + \varepsilon_0} \right) (\varepsilon + \varepsilon_0) = \frac{Q}{\pi a^2} = \sigma .$$

In the limit $a \to 0$ with $Q = \mathrm{const}$ the force acting on the point charge is infinite. Thus in the boundary layer between two dielectrics only surface charges can exist.

## 2.25   Force on a Point Charge by the Field of a Ring Charge in front of a Conducting Sphere

A conducting sphere with radius $a$ and potential $V_0$ is concentrically surrounded by a ring charge $\lambda_q$ in the plane $z = 0$ with radius $b > a$. At a distance $c > a$ from the center of the sphere a point charge $Q$ is located on the $z$-axis. The permittivity $\varepsilon$ is constant.

What is the potential of the sphere, when there is no force acting on the point charge? What is the total charge of the conducting sphere?

For the calculation of the force we need knowledge of the electric field in the outside of the sphere, that can be obtained by means of the method of images for conducting spheres with given potential or with given charge. In this case with given potential we consider a point charge $4\pi\varepsilon a V_0$ in the center ($\varrho = 0$; $z = 0$). Additionally we need image charges of the point charge, $-(a/c)Q$ at position $z = a^2/c$, and of the ring charge $\lambda_q$, which is a line charge $-b/a\lambda_q$ on a circle of radius $\varrho = a^2/b$ in $z = 0$.

The ring charge $\lambda_q$ on the circle $\varrho = b$ in $z = 0$ excites the following potential on the $z$-axis

$$V_1 = \frac{1}{4\pi\varepsilon} \int_s \frac{\lambda_q}{r} ds = \frac{1}{4\pi\varepsilon} \lambda_q \frac{2\pi b}{\sqrt{z^2 + b^2}} = \frac{\lambda_q}{2\varepsilon} \frac{b}{\sqrt{z^2 + b^2}}$$

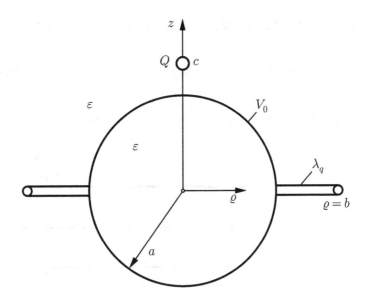

and the electric field becomes

$$\vec{E}_1(\varrho = 0, z) = \vec{e}_z\, E_1(z) = -\vec{e}_z\, \frac{\partial V_1}{\partial z} \; ; \qquad E_1(z) = \frac{\lambda_q}{2\varepsilon}\, \frac{bz}{[z^2 + b^2]^{3/2}} \; .$$

In the same way we get for the image charge $-b/a\lambda_q$

$$E_2(z) = \frac{-\lambda_q b/a}{2\varepsilon}\, \frac{a^2/bz}{[z^2 + (a^2/b)^2]^{3/2}} = -\frac{\lambda_q}{2\varepsilon}\, \frac{az}{[z^2 + (a^2/b)^2]^{3/2}} \; .$$

The charge $-Qa/c$ at position $z = a^2/c$ contributes to the field with

$$E_3(z) = \frac{-Qa/c}{4\pi\varepsilon}\, \frac{1}{(z - a^2/c)^2}$$

and the point charge $4\pi\,\varepsilon\, a\, V_0$ in $(\varrho = 0\,;\, z = 0)$ contributes with $E_4(z) = V_0\, a/z^2$ .

Thus, the total force $\vec{F} = \vec{e}_z F$ acting on the point charge $Q$ in $z = c$ is

$$F = Q\left[\frac{\lambda_q c}{2\varepsilon}\left(\frac{b}{[c^2 + b^2]^{3/2}} - \frac{a}{[c^2 + (a^2/b)^2]^{3/2}}\right) - \frac{Qa/c}{4\pi\varepsilon}\, \frac{1}{(c - a^2/c)^2} + V_0\, a/c^2\right] \; .$$

The force is zero, if the potential takes the value

$$V_0 = \frac{Q}{4\pi\varepsilon}\, \frac{c}{(c - a^2/c)^2} - \frac{\lambda_q c^3}{2\varepsilon}\left(\frac{b/a}{[c^2 + b^2]^{3/2}} - \frac{1}{[c^2 + (a^2/b)^2]^{3/2}}\right) \; .$$

The total charge of the sphere is

$$Q_K = 4\pi\varepsilon\, a\, V_0 - Q\, a/c - 2\pi a\, \lambda_q \; .$$

## 2.26  Boundary Field of a Parallel-Plate Capacitor

Two conducting plates at surfaces $(y = \pm a\,;\, x < 0)$ and with potentials $\pm V_0$ build a parallel-plate capacitor.

Calculate the electric field close to the edges of the plates.

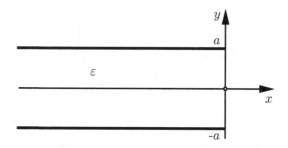

For the calculation of the field one can make use of a conformal transformation in form of a Schwarz-Christoffel-Mapping. As the plane $y = 0$ is a symmetry plane with the potential $V = 0$, it is sufficient to map the half-space $y \geq 0$ with complex coordinates $z = x + jy$ to the half-space $u \geq 0$ with coordinates $w = u + jv$. Furthermore the plates of the capacitor are assumed to be of infinite length in negative $x$-direction.

The drawing below shows the position of three corresponding points in the $z$- and $w$-plane.

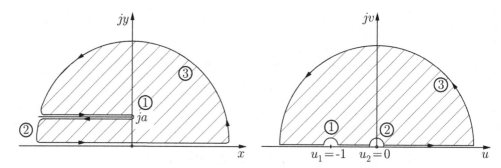

If one mapping point in the $w$-plane is at infinity, then the general mapping function for a polygon with $n$-edges and interior angles $\alpha_\lambda$ is

$$z(w) = \underline{C}_1 \int \prod_{\lambda=1}^{n-1} (w - u_\lambda)^{-\alpha_\lambda} dw + \underline{C}_2.$$

Thus for the given problem the mapping function becomes

$$z(w) = \underline{C}_1 \int (w+1)^1 \, w^{-1} \, dw + \underline{C}_2 = \underline{C}_1[w + \ln(w)] + \underline{C}_2,$$

where the two arbitrary real numbers $u_1$ and $u_2$ have been set to $u_1 = -1$ and $u_2 = 0$. For the determination of the constants $\underline{C}_1$ and $\underline{C}_2$ the correspondence of the first point yields

$$z(w = -1) = ja = \underline{C}_1[-1 + \ln(-1)] + \underline{C}_2 = \underline{C}_1[j\pi - 1] + \underline{C}_2.$$

As it is at infinity in the $z$-plane, the correspondence of the second point requires an integration over the related paths. In the $w$-plane this path is a half circle around $w = 0$ with radius $\varrho \to 0$ and in the $z$-plane the integration is done from $ja - c$ to $-c$ with $c \to \infty$.

$$\int_{ja-c}^{-c} dz = \underline{C}_1 \int_{\varphi=\pi}^{0} \frac{w+1}{w} dw ; \quad w = \varrho \exp(j\varphi) ; \quad dw = \varrho \exp(j\varphi) \, jd\varphi$$

$$-ja = \underline{C}_1 \int_{\varphi=\pi}^{0} \frac{1 + \varrho \exp(j\varphi)}{\varrho \exp(j\varphi)} \varrho \exp(j\varphi) \bigg|_{\varrho \to 0} jd\varphi = \underline{C}_1(-j\pi)$$

$$\underline{C}_1 = \underline{C}_2 = a/\pi ; \quad z(w) = \frac{a}{\pi}[1 + w + \ln(w)]$$

With this, the mapping function is known. The contour in the $z$-plane is mapped onto the real axis in the $w$-plane with the potential $V = V_0$ in $u < 0$ and the potential $V = 0$ in $u > 0$. These potentials excite a field with rotational symmetry in the $w$-plane and thus with equipotential surfaces $\arg(w) = \text{const}$. This potential is that of a virtual line source $\lambda_q$ in $w = 0$, but multiplied with $j$ to swap the real and imaginary part of the complex potential

$$\underline{P}_e(w) = -j \frac{\lambda_q}{2\pi\varepsilon} \ln(w) \quad \text{with} \quad V(u,v) = \text{Re}\{\underline{P}_e(w)\} = \frac{\lambda_q}{\varepsilon} \frac{\varphi}{2\pi}.$$

Now with $V(u > 0, v = 0) = 0$ and $V(u < 0, v = 0) = V_0$ it follows

$$\underline{P}_e(w) = -j \frac{V_0}{\pi} \ln(w).$$

The inverse transformation of the complex potential $\underline{P}_e(w(z)) = \underline{P}_e(z)$ has to be done numerically. Like in many other cases the analytic expression of the inverse function of $w(z)$ is not known.

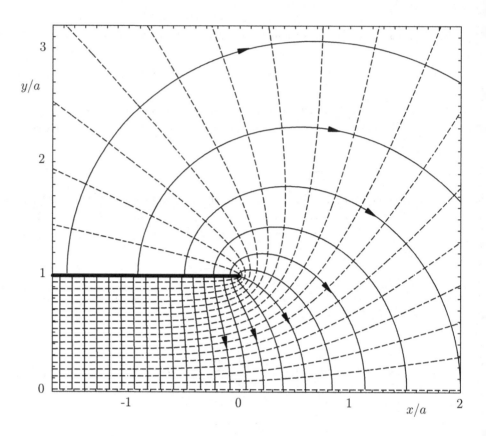

Fig. 2.26–1: Equipotential lines and electric lines of force

The complex electric field $\underline{E}(z) = E_x(x,y) + j\,E_y(x,y)$ is derived from the potential by

$$
\begin{aligned}
\underline{E}(z) &= -\left(\frac{d\underline{P}_e}{dz}\right)^* = -\left(\frac{d\underline{P}_e}{dw}\frac{dw}{dz}\right)^* = -\left(\frac{d\underline{P}_e(w)}{dw}\right)^*\frac{1}{(dz/dw)^*}\\
&= -\left(-j\frac{V_0}{\pi}\frac{1}{w}\right)^*\frac{1}{a/\pi(1+1/w^*)} = -j\frac{V_0}{a}\frac{1}{1+w^*(z)} = E_x + j\,E_y\,.
\end{aligned}
$$

# 3. Stationary Current Distributions

## 3.1  Current Radially Impressed in a Conducting Cylinder

The radial component of the electric field $E_\varrho$ on the surface of a cylinder $\varrho \leq a$ with conductivity $\kappa$ is given by

$$E_\varrho(\varrho = a) = E_0 \begin{cases} \cos(k\varphi) & ; \quad |\varphi| < \pi/3 \\ 0 & ; \quad \pi/3 \leq \varphi \leq 5\pi/3 \end{cases}.$$

For which values of $k$ is this problem well-defined? Calculate the current distribution inside the cylinder $\varrho \leq a$ and the dissipated power per unit length.

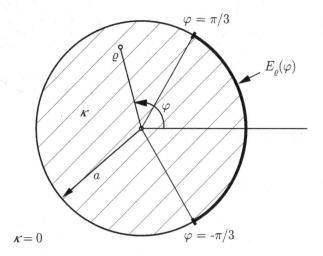

The stationary current density $\vec{J}(\vec{r}) = \kappa \vec{E} = -\kappa \, \mathrm{grad}\, V$ can be described by a potential $V(\vec{r})$ with $\Delta V = 0$. On the surface of the conducting cylinder the normal derivative $\partial V/\partial n$ of the potential is given. As the cylinder is free of sources, the normal derivative must fulfill

$$\int_{\varphi=0}^{2\pi} \frac{\partial V}{\partial n}\bigg|_{\varrho=a} a\, d\varphi = 0 = \int_0^{2\pi} \frac{\partial V}{\partial \varrho}\bigg|_{\varrho=a} a\, d\varphi = -E_0 \int_{\varphi=-\pi/3}^{\pi/3} \cos(k\varphi)\, a\, d\varphi = 0.$$

This condition holds for $k = 3p$ with $p = 1, 2, 3, \ldots$.

The approach for the potential is

$$V(\varrho, \varphi) = A_0 + \sum_{n=1}^{\infty} A_n (\varrho/a)^n \cos(n\varphi) ; \qquad \varrho \leq a ,$$

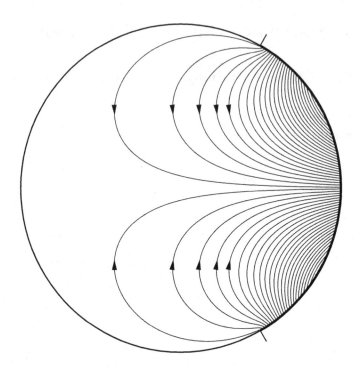

Fig. 3.1–1: Current streamlines for $k = 3$

which has to satisfy the boundary condition

$$E_\varrho\big|_{\varrho=a} = -\frac{\partial V}{\partial \varrho}\bigg|_{\varrho=a} = -\sum_{n=1}^{\infty} nA_n/a\cos(n\varphi) = E_0 \begin{cases} \cos(3p\varphi) & ; \quad |\varphi| < \pi/3 \\ 0 & ; \quad \pi/3 \le \varphi \le 5\pi/3 \end{cases}.$$

Making use of the orthogonality relation for trigonometric functions by multiplication with $\cos(m\varphi)$ and integration leads to

$$-m\pi\, A_m/a = E_0 \int_{-\pi/3}^{\pi/3} \cos(3p\varphi)\cos(m\varphi)d\varphi$$

$$A_n = -\frac{aE_0}{n\pi}\begin{cases} \pi/3 & ; \quad n = 3p \\ \dfrac{\sin((3p-n)\pi/3)}{3p-n} + \dfrac{\sin((3p+n)\pi/3)}{3p+n} & ; \quad n \ne 3p \end{cases}.$$

The constant $A_0$ remains unknown for boundary value problems of this kind.

The current density is

$$\vec{J} = -\frac{\kappa}{a} \sum_{n=1}^{\infty} n A_n (\varrho/a)^{n-1} (\vec{e}_\varrho \cos(n\varphi) - \vec{e}_\varphi \sin(n\varphi))$$

and the dissipated power in the cylinder per length $l$ is

$$P_v/l = \int_v \vec{J}\vec{E}\, dv/l =$$

$$= \frac{1}{\kappa} \left(\frac{\kappa}{a}\right)^2 \sum_{n=1}^{\infty} n^2 A_n^2 \int_{\varrho=0}^{a} (\varrho/a)^{2n-2} \varrho\, d\varrho \int_{\varphi=0}^{2\pi} (\cos^2(n\varphi) + \sin^2(n\varphi)) d\varphi$$

$$= \kappa \pi \sum_{n=1}^{\infty} n A_n^2 \,.$$

## 3.2   Current Distribution around a Hollow Sphere

Consider a homogeneous conducting space with conductivity $\kappa$. The initial current density $\vec{J_0} = -\vec{e}_z J_0$ is either homogeneous. How does the current distribution change, if a spherical region of radius $a$ is excluded and filled with non-conducting material of permittivity $\varepsilon_0$.

Calculate the potential and the electric field inside and outside the sphere. What is the charge density on the surface of the sphere?

The initial homogeneous current density

$$\vec{J_0} = -\vec{e}_z J_0 = -\vec{e}_z \kappa E_0 = -\kappa \operatorname{grad} V_0 = -\vec{e}_z \kappa \frac{\partial V_0}{\partial z}$$

is described by the potential $V_0 = E_0 z = E_0 r \cos \vartheta$.

The potential of the current problem with a non-conducting sphere

$$V = V_0 + V_s \quad \text{with} \quad \Delta V_s = 0$$

is then given by

$$V = \begin{cases} V_0 + \sum\limits_{n=0}^{\infty} A_n (a/r)^{n+1} P_n(\cos \vartheta) \quad ; \quad r \geq a \\[2mm] \sum\limits_{n=0}^{\infty} B_n (r/a)^n P_n(\cos \vartheta) \qquad\quad ; \quad r \leq a \end{cases}.$$

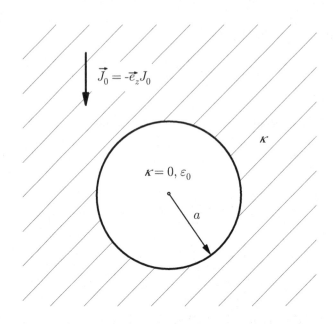

As the potential of the initial field only includes the Legendre polynomial $P_1 = \cos \vartheta$

$$V_0 = E_0 \, a \left( \frac{r}{a} \right) P_1(\cos \vartheta),$$

we can leave out all other polynomials in our solution, thus:

$$V(r,\vartheta) = \begin{cases} V_0 + A \, (a/r)^2 \cos \vartheta & ; \quad r \geq a \\ B \, r/a \cos \vartheta & ; \quad r \leq a \end{cases}.$$

At the boundary to the non-conducting sphere the current streamlines are tangential. Thus the normal derivative of the potential vanishes at $r = a$.

$$\left. \frac{\partial V}{\partial r} \right|_{r>a, r\to a} = (E_0 - 2A/a) \cos \vartheta = 0 \to A = \frac{a \, E_0}{2},$$

The potential itself is continuous.

$$E_0 \, a + A = B \, ; \qquad B = \frac{3}{2} a \, E_0$$

Now the result for the potential is

$$V(r,\vartheta) = a \, E_0 \begin{cases} r/a + 1/2 \, (a/r)^2 \\ 3/2 \, r/a \end{cases} \cos \vartheta \, ; \qquad \begin{matrix} r \geq a \\ r \leq a \end{matrix}.$$

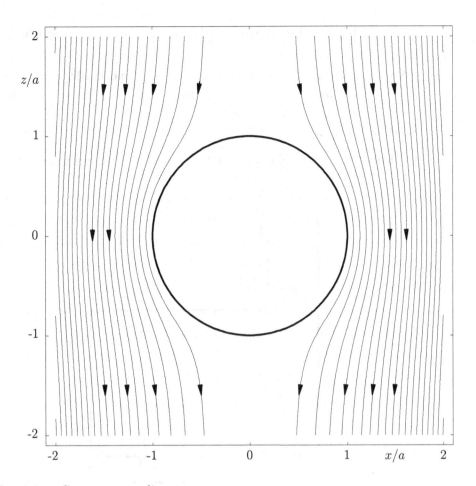

Fig. 3.2–1: Current streamlines

Inside the sphere $r < a$ the electric field becomes

$$\vec{E} \;=\; -\operatorname{grad} V|_{r<a} \;=\; -\vec{e}_z \frac{\partial}{\partial z} E_0 \frac{3}{2} (r \cos \vartheta) \;=\; -\vec{e}_z \frac{3}{2} E_0 \,.$$

Thus in $r < a$ the electric field remains homogeneous.

The charge density at $r = a$ results from the normal derivative of the potential $V|_{r<a}$

$$\sigma \;=\; -\varepsilon_0 \left.\frac{\partial V}{\partial n}\right|_{r=a} \;=\; \varepsilon_0 \left.\frac{\partial V}{\partial r}\right|_{r=a} \;=\; \varepsilon_0 \frac{3}{2} E_0 \cos \vartheta \,.$$

Finally the current density $\vec{J} = -\kappa \operatorname{grad} V$ in $r > a$ is

$$
\begin{aligned}
\vec{J} \;&=\; -\kappa\, E_0 \left[ (\vec{e}_r \cos \vartheta - \vec{e}_\vartheta \sin \vartheta) - (a/r)^3 (\vec{e}_r \cos \vartheta + 1/2\,\vec{e}_\vartheta \sin \vartheta) \right] \\
&=\; -J_0 \left[ \vec{e}_z - (a/r)^3 (\vec{e}_r \cos \vartheta + 1/2\,\vec{e}_\vartheta \sin \vartheta) \right] \,.
\end{aligned}
$$

## 3.3   Current Distribution inside a Rectangular Cylinder

A rectangular cylinder with cross section $a \cdot b$ and conductivity $\kappa$ is connected to perfectly conducting electrodes of width $2s \ll b < a$ at the top and the bottom. The total current flow per axial length is $I/l$ and the field is assumed to be independent of the $z$-coordinate.

Calculate the current distribution, the power loss $P_v/l$, and the resistance $R/l = P_v/(lI^2)$.

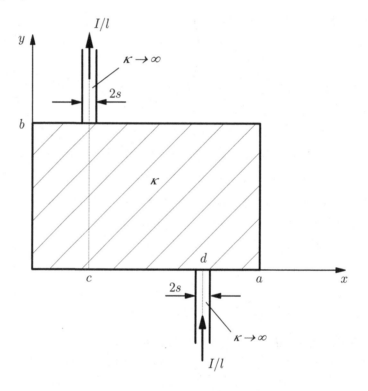

On the boundary of the cylinder the normal derivation of the potential $\partial V/\partial n = 0$ vanishes, except for the range of the conducting electrodes, where the potential is constant because of the perfect conductivity. Hence the problem is a mixed boundary value problem. However, as $2s \ll b$ we can assume, that the normal derivative of the potential at the ends of the electrodes is constant. The normal component of the current density on the boundary $y = b$ becomes

$$J_n = \frac{I/l}{2s} = -\kappa \frac{\partial V}{\partial n} = -\kappa \frac{\partial V}{\partial y} \; ; \qquad \left. \frac{\partial V}{\partial y} \right|_{y=b} = -\frac{I/l}{2\kappa s} \; ; \qquad |x - c| < s$$

and on $y = 0$

$$J_n = -\frac{I/l}{2s} = -\kappa\frac{\partial V}{\partial n} = \kappa\frac{\partial V}{\partial y} \; ; \qquad \frac{\partial V}{\partial y}\bigg|_{y=0} = -\frac{I/l}{2\kappa s} \; ; \qquad |x - d| < s \, .$$

The boundary conditions for the normal derivative of the potential on $x = 0$ and $x = b$ are homogeneous. This is not the case for the planes $y = 0$ and $y = b$. As with rot $\vec{E} = 0$; div $\vec{J} = 0$, and $\vec{J} = \kappa\vec{E}$ the potential $V$ is a solution of $\Delta V = 0$, a practical approach for the current problem is

$$V(x, y) = A_0 + B_0 y + \sum_{n=1}^{\infty} \frac{\cos(n\pi x/a)}{\sinh(n\pi b/a)} \left[ A_n \cosh(n\pi y/a) + B_n \cosh(n\pi(y-b)/a) \right] .$$

The homogeneous boundary conditions $\partial V/\partial x|_{x=0,a} = 0$ are already satisfied and for the normal derivative $\partial V/\partial y$ in $y = b$ it follows

$$\frac{\partial V}{\partial y}\bigg|_{y=b} = B_0 + \sum_{n=1}^{\infty} \cos(n\pi x/a)\,(n\pi/a)\,A_n = \begin{cases} -I/(2\kappa s l) & ; \quad |x - c| < s \\ 0 & ; \quad |x - c| \geq s. \end{cases}$$

Making use of the orthogonality of the trigonometric functions results after multiplication with $\cos(m\pi x/a)$ and integration in

$$B_0 a = -\frac{I/l}{2\kappa s}\, 2s \; ; \qquad A_m\,(m\pi/a)\,a/2 = -\frac{I/l}{2\kappa s} \int\limits_{c-s}^{c+s} \cos(m\pi x/a)dx$$

$$B_0 = -\frac{I/l}{\kappa a} \; ; \qquad A_n = -\frac{2I/l}{\kappa n\pi}\cos(n\pi c/a)\,\frac{\sin(n\pi s/a)}{n\pi s/a} \, .$$

Doing the same for the boundary $y = 0$ leads to

$$\frac{\partial V}{\partial y}\bigg|_{y=0} = B_0 - \sum_{n=1}^{\infty} \cos(n\pi x/a)\,B_n\,(n\pi/a) = \begin{cases} -I/(2\kappa s l) & ; \quad |x - d| < s \\ 0 & ; \quad |x - d| \geq s \end{cases}$$

$$B_0 = -\frac{I/l}{\kappa a} \; ; \qquad B_n = \frac{2I/l}{\kappa n\pi}\cos(n\pi d/a)\,\frac{\sin(n\pi s/a)}{n\pi s/a} \, .$$

The result for $B_0$ is the same as before, as the total current is constant and thus the problem is well-defined. The constant $A_0$ remains undetermined.

The calculation of the power loss per axial length $P_v/l$ requires the solution of

$$P_v/l = \int\limits_{y=0}^{b} \int\limits_{x=0}^{a} \vec{J}\vec{E}dx\,dy = \kappa \int\limits_{y=0}^{b} \int\limits_{x=0}^{a} (\operatorname{grad} V)^2\,dx\,dy \, .$$

A straight forward evaluation of this surface integrals is possible, but with Green's second identity the surface integral is transformed into a simpler integral over the contour $C$ of the conducting cylinder.

$$P_v/l = \kappa \oint_C V \frac{\partial V}{\partial n} ds = \kappa V(x = c, y = b) \int_{c-s}^{c+s} \frac{\partial V}{\partial n} dx + \kappa V(x = d, y = 0) \int_{d-s}^{d+s} \frac{\partial V}{\partial n} dx$$

$$P_v/l = V(x = c, y = b) \underbrace{\int_{c-s}^{c+s} \kappa \frac{\partial V}{\partial y} dx}_{-I/l} + V(x = d, y = 0) \underbrace{\int_{d-s}^{d+s} \left(-\kappa \frac{\partial V}{\partial y}\right) dx}_{I/l}$$

This result leads to

$$P_v/l = I/l \left[ V(x = d, y = 0) - V(x = c, y = b) \right] = I/l \cdot U = I^2 R/l$$

$$P_v/l = I/l \left[ -B_0 b + \sum_{n=1}^{\infty} \frac{\cos(n\pi d/a)}{\sinh(n\pi b/a)} \left( A_n + B_n \cosh(n\pi b/a) \right) + \right.$$
$$\left. - \frac{\cos(n\pi c/a)}{\sinh(n\pi b/a)} \left( A_n \cosh(n\pi b/a) + B_n \right) \right] .$$

In the limit $s \to 0$ the supply current is impressed on a line. With $\lim\limits_{x \to 0} \sin(x)/x = 1$ the constants in the series for the potential and current density are

$$A_n|_{s \to 0} = -\frac{2I/l}{\kappa n\pi} \cos(n\pi c/a) ; \qquad B_n|_{s \to 0} = \frac{2I/l}{\kappa n\pi} \cos(n\pi d/a).$$

However, the series for the power loss and the resistance are divergent, i.e. with $s \to 0$ the resistance increases unrestricted.

## 3.4  Current Distribution inside a Circular Cylinder

This problem resembles problem 3.3, but the rectangular cylinder has been replaced by a circular cylinder with conductivity $\kappa$ and radius $a$. The input current supply lies in the half plane $\varphi = \varphi_0$ and the output current supply in $\varphi = 0$.

Calculate the current distribution and the resistance per axial length. In the limit case $c \to 0$ it is possible to calculate the current distribution with the method of images. Calculate the potential by means of equivalent line sources.

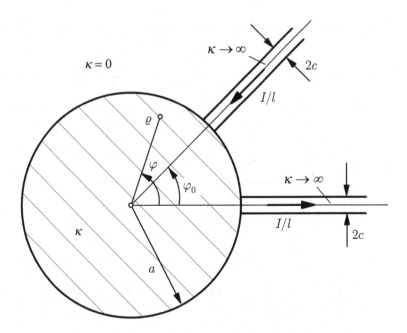

The potential on the surface of the cylinder is constant in the region of the perfect conducting electrodes. Furthermore because of $2c \ll a$ we can assume that the normal component of the current density is either constant at the end of the electrodes. Thus the constant normal derivative of the potential is

$$-\frac{\partial V}{\partial n}\bigg|_{\varrho=a} = -\frac{\partial V}{\partial \varrho}\bigg|_{\varrho=a} = E_n(\varphi)|_{\varrho=a} = E_\varrho(\varphi)|_{\varrho=a} =$$

$$= \begin{cases} \dfrac{I}{2c\kappa l} & ; & -\dfrac{c}{a} < \varphi < \dfrac{c}{a} \\[2mm] -\dfrac{I}{2c\kappa l} & ; & \varphi_0 - \dfrac{c}{a} < \varphi < \varphi_0 + \dfrac{c}{a} \\[2mm] 0 & ; & \text{otherwise} \end{cases}$$

$$\text{with} \quad \int\limits_0^{2\pi} \frac{\partial V}{\partial n}\bigg|_{\varrho=a} a\, d\varphi = 0 \,.$$

On account of the symmetry plane $\varphi = \varphi_0/2$ the approach for the potential $V(\varrho,\varphi)$ reads

$$V(\varrho,\varphi) = \sum_{n=1}^{\infty} C_n \left(\frac{\varrho}{a}\right)^n \sin\big(n(\varphi - \varphi_0/2)\big).$$

The term with $n = 0$ can be omitted because the point $\varrho = 0$ is part of the computation domain. For the calculation of the constants $C_n$ we look at the normal component of the electric field at points on the cylinder $\varrho = a$.

$$E_\varrho\big|_{\varrho=a} = -\frac{\partial V}{\partial \varrho}\bigg|_{\varrho=a} = -\sum_{n=1}^{\infty} C_n \frac{n}{a} \sin\big(n(\varphi - \varphi_0/2)\big)$$

Making use of the orthogonality relation for trigonometric functions results in

$$-\sum_{n=1}^{\infty} C_n \frac{n}{a} \underbrace{\int_0^{2\pi} \sin\big(n(\varphi - \varphi_0/2)\big) \sin\big(k(\varphi - \varphi_0/2)\big)d\varphi}_{= \begin{cases} \pi \; ; \; n = k \\ 0 \; ; \; n \neq k \end{cases}} = \int_0^{2\pi} E_\varrho(\varphi)\big|_{\varrho=a} \sin\big(k\,(\varphi-\varphi_0/2)\big)d\varphi$$

$$-C_n \frac{n}{a}\pi = \int_{-c/a}^{c/a} \frac{I}{2c\kappa l} \sin\big(n\,(\varphi - \varphi_0/2)\big)d\varphi + \int_{\varphi_0-\frac{c}{a}}^{\varphi_0+\frac{c}{a}} \frac{-I}{2c\kappa l} \sin\big(n\,(\varphi - \varphi_0/2)\big)d\varphi$$

$$= \frac{I}{2c\kappa l} \int_{-c/a}^{c/a} \big[\sin\big(n\,(\varphi - \varphi_0/2)\big) - \sin\big(n\,(\varphi + \varphi_0/2)\big)\big]\,d\varphi$$

$$= \frac{-I}{c\kappa l} \sin(n\,\varphi_0/2) \int_{-c/a}^{c/a} \cos\,(n\,\varphi)\,d\varphi = -\frac{2I}{c\kappa l} \sin\,(n\,\varphi_0/2)\frac{\sin\,(n\,c/a)}{n}\,.$$

Finally the result is

$$C_n = \frac{2I}{\kappa\pi l}\frac{\sin\,(n\,\varphi_0/2)}{n}\frac{\sin\,(n\,c/a)}{n\,c/a}$$

$$V(\varrho,\varphi) = \frac{2I}{\kappa\pi l}\sum_{n=1}^{\infty}\frac{\sin\,(n\varphi_0/2)}{n}\frac{\sin\,(n\,c/a)}{n\,c/a}\Big(\frac{\varrho}{a}\Big)^n \sin\big(n\,(\varphi - \varphi_0/2)\big)\,.$$

The resistance per axial length $R/l$ is defined by the relation

$$\frac{1}{l}\frac{V(a,\varphi_0) - V(a,0)}{I} = \frac{R}{l} = \frac{4}{\kappa\pi l^2}\sum_{n=1}^{\infty}\frac{\sin^2\,(n\,\varphi_0/2)}{n}\frac{\sin\,(n\,c/a)}{n\,c/a}$$

with $\quad V(a,0) = -V(a,\varphi_0)\,.$

In the limit case $c/a \to 0$ with $\lim\limits_{x\to 0}\sin(x)/x = 1$ the potential is

$$V_0(\varrho,\varphi) = \frac{2I}{\kappa\pi l}\sum_{n=1}^{\infty}\frac{\sin\,(n\varphi_0/2)}{n}\Big(\frac{\varrho}{a}\Big)^n \sin\big(n\,(\varphi - \varphi_0/2)\big)\,.$$

The resistance for $\varphi_0 = \pi$ and $c/a \to 0$ reads:

$$\frac{R}{l} = \frac{4}{\kappa \pi l^2} \sum_{k=0}^{\infty} \frac{1}{2k+1}.$$

This series is divergent. The potentials for the points $(\varrho = a; \varphi = \varphi_0)$ and $(\varrho = a; \varphi = 0)$ and the resistance $R/l$ increase unrestricted for $c/a \to 0$.

This result can also be derived by means of the method of images, in analogy to the method for line charges in front of a dielectric cylinder. For line currents $\pm I$ on the surface of a conducting cylinder one gets image currents at the same positions, thus in total line currents $\pm 2I$ at positions $(\varrho = a; \varphi = \varphi_0)$ and $(\varrho = a; \varphi = 0)$ in a space of homogeneous conductivity $\kappa$. Now the resulting potential $V(\varrho, \varphi)$ is the real part of the complex potential $\underline{P}(z)$.

$$\underline{P}(z) = -\frac{I}{\pi \kappa l} \ln \frac{z - z_1}{z - z_2}; \qquad z_1 = a; \qquad z_2 = a \exp(j\varphi_0)$$

$$V(\varrho, \varphi) = \operatorname{Re}\{\underline{P}(z)\} = -\frac{I}{\pi \kappa l} \ln(\varrho_1/\varrho_2); \qquad \varrho_i = |z - z_i|; \qquad i = 1, 2$$

The complex electric field is

$$\underline{E}(z) = -(d\underline{P}/dz)^* = \frac{I}{\pi \kappa l} [1/(z - z_1) - 1/(z - z_2)]^*$$

and the lines of force are defined by the electric stream function

$$\Psi_s = \operatorname{Im}\{\underline{P}(z)\} = \text{const}.$$

Although the resulting solution from the method of images is of different form it is identical with the preceding result. For the proof we look at a series expansion of the logarithm of the distance $|z - z_0| = \varrho_0$.

$$z = \varrho \exp(j\varphi); \qquad z_0 = |z_0| \exp(j\varphi_0); \qquad |z_0| = a$$

$$\ln \varrho_0 = \ln a - \sum_{n=1}^{\infty} (\varrho/a)^n/n \, \cos(n(\varphi - \varphi_0)); \qquad \varrho < a$$

Now we can rewrite the real part of the complex potential

$$V(\varrho, \varphi) = -\frac{I}{\kappa \pi l} \sum_{n=1}^{\infty} (\varrho/a)^n/n \cdot \underbrace{[\cos(n\varphi) - \cos(n(\varphi - \varphi_0))]}_{-2 \sin(n(\varphi - \varphi_0 + \varphi)/2) \sin(n(\varphi + \varphi_0 - \varphi)/2)}$$

$$V(\varrho, \varphi) = \frac{2I}{\kappa \pi l} \sum_{n=1}^{\infty} (\varrho/a)^n/n \sin(n\varphi_0/2) \sin(n(\varphi - \varphi_0/2))$$

and get the same expression as before.

## 3.5 Current Distribution in a Cylinder with Stepped Down Diameter

Consider a circular conductor of infinite extension in $\pm z$-direction that carries the current $I$. The radius of the conductor reduces in $z = 0$ from $b$ to $a$, with $a < b$.

Calculate the current distribution in the conductor of conductivity $\kappa$.

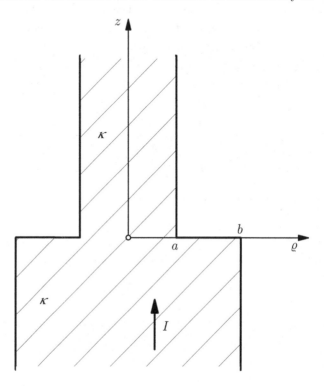

The current distribution far away from $z = 0$ is homogeneous.

$$\vec{e}_z\ \vec{J}(\vec{r})\Big|_{z \to \infty} = J_z\big|_{z \to \infty} = \frac{I}{\pi a^2}$$

That is

$$E_z\big|_{z \to \infty} = \frac{I}{\kappa \pi a^2} = -\frac{\partial V}{\partial z}\Big|_{z \to \infty} \quad ; \qquad \frac{\partial V}{\partial z}\Big|_{z \to \infty} = -E_0$$

and in analogy

$$\frac{\partial V}{\partial z}\Big|_{z \to -\infty} = -E_0 \left(\frac{a}{b}\right)^2 .$$

We thus need linear functions in $z$ with $|z| \to \infty$ for the description of the potential. Furthermore the solution functions must satisfy Laplace's equation. This leads to the following approach

$$V_1(\varrho, z) = -E_0 z + \sum_{m_1} A_{m_1} J_0(m_1\varrho) \exp(-m_1 z); \quad z \geq 0$$

$$V_2(\varrho, z) = -E_0 \left(\frac{a}{b}\right)^2 z + \sum_{m_2} B_{m_2} J_0(m_2\varrho) \exp(m_2 z); \quad z \leq 0.$$

The radial component of the current density vanishes on the boundary of the cylinder. This leads for $\varrho = a$ and $\varrho = b$ with $J_0' = -J_1$ to

$$\frac{\partial V_1}{\partial \varrho}\bigg|_{\varrho=a} = 0 \quad \Rightarrow \quad J_1(m_1 a) = 0; \qquad \frac{\partial V_2}{\partial \varrho}\bigg|_{\varrho=b} = 0 \quad \Rightarrow \quad J_1(m_2 b) = 0$$

$$J_1(x_{1r}) = 0; \quad r = 1, 2, 3, \ldots \quad \Rightarrow \quad m_{1r} = \frac{x_{1r}}{a}; \qquad m_{2r} = \frac{x_{1r}}{b}$$

$$V_1(\varrho, z) = -E_0 z + \sum_{r=1}^{\infty} A_r J_0(x_{1r}\varrho/a) \exp(-x_{1r}z/a); \quad z \geq 0$$

$$V_2(\varrho, z) = -E_0 \left(\frac{a}{b}\right)^2 z + \sum_{r=1}^{\infty} B_r J_0(x_{1r}\varrho/b) \exp(x_{1r}z/b); \quad z \leq 0.$$

The remaining constants $A_r$ and $B_r$ follow from the boundary conditions in $z = 0$. The evaluation of the condition for the normal component of the current density gives

$$E_{z2}\big|_{z=0} = \begin{cases} E_{z1}\big|_{z=0} ; & \varrho < a \\ 0 & ; \quad a < \varrho < b \end{cases}$$

$$E_0 \left(\frac{a}{b}\right)^2 - \sum_{r=1}^{\infty} B_r \, x_{1r}/b \, J_0\left(x_{1r}\,\varrho/b\right) = \begin{cases} E_0 + \sum_{r=1}^{\infty} A_r \, x_{1r}/a \, J_0(x_{1r}\varrho/a) ; & \varrho < a \\ 0 & ; \quad a < \varrho < b. \end{cases}$$

As the functions $J_0(x_{1r}\varrho/b)$ belong to an orthogonal system of functions, the multiplication with $\varrho J_0(x_{1s}\varrho/b)$ and integration from $\varrho = 0$ to $\varrho = b$ results in

$$E_0 \left(\frac{a}{b}\right)^2 \underbrace{\int_0^b J_0(x_{1s}\frac{\varrho}{b}) \varrho d\varrho}_{I_1} - \sum_{r=1}^{\infty} B_r \frac{x_{1r}}{b} \underbrace{\int_0^b J_0(x_{1r}\frac{\varrho}{b}) J_0(x_{1s}\frac{\varrho}{b}) \varrho d\varrho}_{I_2} =$$

$$= E_0 \underbrace{\int_0^a J_0(x_{1s}\varrho/b) \varrho d\varrho}_{I_3} + \sum_{r=1}^{\infty} A_r \frac{x_{1r}}{a} \underbrace{\int_0^a J_0(x_{1r}\varrho/a) J_0(x_{1s}\varrho/b) \varrho d\varrho}_{I_4}$$

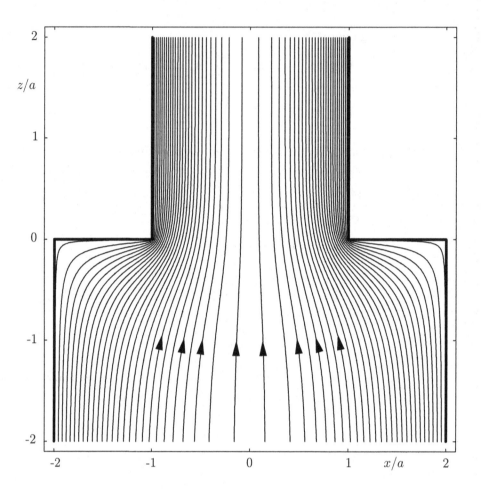

Fig. 3.5–1: Current streamlines

with

$$I_1 = 0 \; ; \qquad I_2 = \frac{1}{2} \, b^2 \, J_0^2(x_{1s}) \; ; \qquad I_3 = a \, \frac{b}{x_{1s}} \, J_1(x_{1s}a/b)$$

$$I_4 = -\frac{x_{1s} \, a/b \, J_1(x_{1s}a/b) \, J_0(x_{1r})}{(x_{1r}/a)^2 - (x_{1s}/b)^2} \; .$$

This leads to the following system of equations

$$\frac{1}{2} \, B_s \, \frac{J_0^2(x_{1s})}{J_1(x_{1s}\,a/b)} = -E_0 \, \frac{a}{x_{1s}^2} + \sum_{r=1}^{\infty} A_r \, \frac{x_{1r} \, J_0(x_{1r})}{(x_{1r}\,b/a)^2 - x_{1s}^2} \; .$$

Finally the boundary condition for the tangential component of the electric field, which is equivalent to the continuity of the potential, is evaluated in a similar manner.

$$\left.\frac{\partial V_1}{\partial \varrho}\right|_{z=0} = \left.\frac{\partial V_2}{\partial \varrho}\right|_{z=0} \quad \Leftrightarrow \quad V_1(\varrho, z=0) = V_2(\varrho, z=0); \quad \varrho < a$$

$$\Rightarrow \quad \sum_{r=1}^{\infty} A_r J_0(x_{1r}\varrho/a) = \sum_{r=1}^{\infty} B_r J_0(x_{1r}\varrho/b); \quad \varrho < a.$$

Now the multiplication with $\varrho J_0(x_{1s}\,\varrho/a)$ and integration from $\varrho = 0$ to $\varrho = a$ results in

$$A_s \frac{1}{2} a^2 J_0^2(x_{1s}) = \sum_{r=1}^{\infty} B_r \underbrace{\int_0^a J_0(x_{1r}\,\varrho/b)\, J_0(x_{1s}\,\varrho/a)\, \varrho d\varrho}_{\displaystyle \frac{x_{1r}\,a/b\,J_1(x_{1r}a/b)\,J_0(x_{1s})}{(x_{1r}/b)^2 - (x_{1s}/a)^2}}$$

and leads to a second system of equations.

$$A_s = \frac{2}{J_0(x_{1s})} \sum_{r=1}^{\infty} B_r \frac{x_{1r}\,a/b\,J_1(x_{1r}\,a/b)}{(x_{1r}\,a/b)^2 - x_{1s}^2}$$

The electric field and thus the current distribution is determined after solving both systems of equations numerically.

## 3.6 Current Distribution around a Conducting Sphere

The initial current density $\vec{J}_E = -\vec{e}_z J_E$ in the homogeneous space of conductivity $\kappa_2$ is perturbed by a conducting sphere of radius $a$ and conductivity $\kappa_1$.

Calculate the perturbed current distribution.

The initial current density in the homogeneous space of conductivity $\kappa_2$ is

$$\vec{J}_E = -\vec{e}_z\, J_E = -\left[\vec{e}_r \cos\vartheta - \vec{e}_\vartheta \sin\vartheta\right] J_E = \kappa_2 \vec{E}_E.$$

It is described by the Potential $V_E$:

$$\vec{J}_E = -\kappa_2 \mathrm{grad}\, V_E = -\kappa_2 \left[\vec{e}_r \frac{\partial V_E}{\partial r} + \frac{\vec{e}_\vartheta}{r}\frac{\partial V_E}{\partial \vartheta}\right]$$

$$\kappa_2 \frac{\partial V_E}{\partial r} = J_E \cos\vartheta; \qquad \frac{\kappa_2}{r}\frac{\partial V_E}{\partial \vartheta} = -J_E \sin\vartheta.$$

Hence the potential is

$$V_E(r,\vartheta) = \frac{J_E}{\kappa_2}\, r \cos\vartheta = \frac{J_E}{\kappa_2}\, r\, P_1(\cos\vartheta),$$

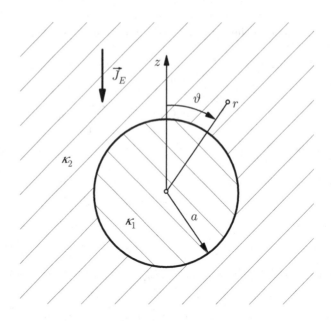

with the definition $V(r, \vartheta = \pi/2) = 0$ without loss of generality.

The additional functions that describe the perturbation by the inserted conducting sphere must be solutions of Laplace's equation $\Delta V(r, \vartheta) = 0$. In spherical coordinates these are the powers of the distance $(r^n, r^{-(n+1)})$ and the Legendre's polynomials $P_n(\cos \vartheta)$ and $Q_n(\cos \vartheta)$. Since the description of the initial field restricts to functions of order $n = 1$, also the perturbed field is described by these functions and the approach for the resulting potential reads as follows

$$V(r, \vartheta) = \frac{J_E\, a}{\kappa_2} \left\{ \begin{array}{c} r/a + C_1 (a/r)^2 \\ C_2\, r/a \end{array} \right\} P_1(\cos \vartheta)\,; \qquad \begin{array}{c} r \geq a \\ r \leq a \end{array}\,.$$

The unknown constants $C_1$ and $C_2$ result from the boundary conditions in the surface $r = a$.

The continuity of the potential requires

$$1 + C_1 = C_2.$$

A second equation follows from the continuity of the normal component of the current density

$$\kappa_2\, \left.\frac{\partial V}{\partial r}\right|_{\substack{r > a \\ r \to a}} = \kappa_1\, \left.\frac{\partial V}{\partial r}\right|_{\substack{r < a \\ r \to a}} \qquad \Rightarrow \qquad 1 - 2\,C_1 = \kappa_1/\kappa_2\, C_2\,.$$

Accordingly the constants are

$$C_1 = \frac{\kappa_2 - \kappa_1}{\kappa_1 + 2\kappa_2} \; ; \qquad C_2 = \frac{3\kappa_2}{\kappa_1 + 2\kappa_2}$$

and after building the gradient of the resulting potential the current density becomes

$$\vec{J} = -\kappa \operatorname{grad} V = J_E \begin{cases} -\vec{e}_z + \dfrac{\kappa_2 - \kappa_1}{\kappa_1 + 2\kappa_2}(a/r)^3 [2\vec{e}_r \cos \vartheta + \vec{e}_\vartheta \sin \vartheta] \quad ; \quad r > a \\[3mm] -\dfrac{3\kappa_1}{\kappa_1 + 2\kappa_2}\vec{e}_z \qquad\qquad\qquad\qquad\qquad ; \quad r < a. \end{cases}$$

Hence only inside the sphere $r < a$ the current distribution is homogeneous. In the limit case $\kappa_1 \to 0$ we get the same problem and thus the same solution as in chapter 3.2.

For the drawing of current streamlines one can make use of the relation

$$I(r, \vartheta) = \int_a \vec{J}d\vec{a} = 2\pi \int_{\vartheta'=0}^{\vartheta} \vec{e}_r \vec{J}\, r^2 \sin \vartheta' d\vartheta' = \text{const} .$$

The result is

$$I(r, \vartheta) = -2\pi \kappa_{1,2}\, r^2 \int_{\vartheta'=0}^{\vartheta} \frac{J_E a}{\kappa_2 a} \begin{cases} 1 - 2\,C_1(a/r)^3 \\[2mm] C_2 \end{cases} \cos \vartheta' \sin \vartheta' d\vartheta'$$

$$I(r, \vartheta)/I_0 = \begin{cases} 1 - 2\,C_1(a/r)^3 \\[2mm] \kappa_1/\kappa_2 C_2 \end{cases} [\cos 2\vartheta - 1] = \text{const}$$

with $I_0 = \pi r^2 J_E/2$.

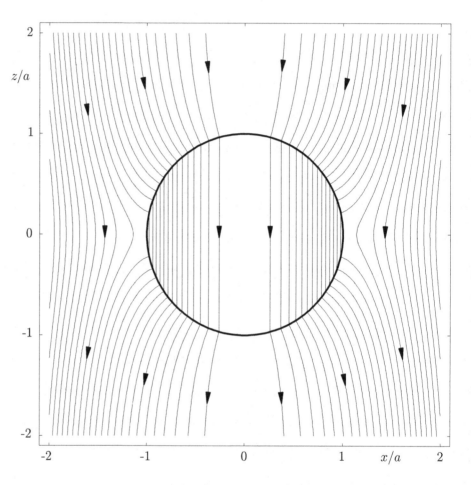

Fig. 3.6–1: Current streamlines for $\kappa_1/\kappa_2 = 100$

# 4. Magnetic Field of Stationary Currents

## 4.1  Magnetic Field of Line Conductors

In the plane $z = 0$ lies a circular conductor loop with radius $a$ and with its center in the point of origin. For the current supply radial line currents have been connected at the points $(\varrho = a \,;\, \varphi = \mp\varphi_0)$, starting at infinity and carrying the constant current $I$.

Calculate the magnetic field at the point of origin $(\varrho = 0 \,;\, z = 0)$.

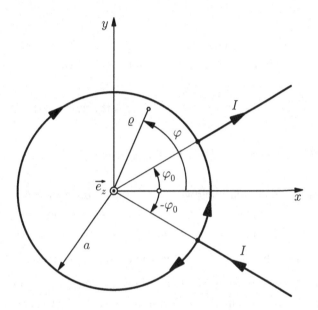

As the current $I$ is time independent, he splits up accordingly to the resistances, which are proportional to the lengths of the wire elements. The element $|\varphi| < \varphi_0$ carries the current $I(1 - \varphi_0/\pi)$ and the element $\varphi_0 < \varphi < 2\pi - \varphi_0$ the current $I\varphi_0/\pi$.

In general the magnetic field of a line current with contour $C(\vec{r}_q)$ is given by

$$\vec{H}(\vec{r}_p) = \frac{I}{4\pi} \oint_C d\vec{s}_q \times \vec{r}/r^3 \;; \qquad \vec{r} = \vec{r}_p - \vec{r}_q \,.$$

Hence line currents with radial direction don't contribute to the field at the point of origin $\vec{r}_p = 0$, because the vectors $\vec{r} = \vec{r}_p - \vec{r}_q = -\vec{r}_q$ and $d\vec{s}_q$ have the same direction and therefore the cross product vanishes. The result of the integral over the contour of the circular element $(\varrho = a \,; -\varphi_0 < \varphi < \varphi_0)$ is

$$I\frac{1-\varphi_0/\pi}{4\pi} \int\limits_{\varphi=-\varphi_0}^{\varphi_0} \vec{e}_\varphi\, a\, d\varphi_q \times (-\vec{e}_\varrho/a^2) = \vec{e}_z \frac{I}{2\pi a}\, (1-\varphi_0/\pi)\, \varphi_0$$

and the result for the complementary element $(\varrho = a \,; \varphi_0 < \varphi < 2\pi - \varphi_0)$ is

$$-\vec{e}_z \frac{I}{2\pi a} \varphi_0(1-\varphi_0/\pi).$$

Thus the magnetic field in the point of origin vanishes.

$$\vec{H} = 0$$

If the current $I$ would split-up equally onto the wire elements, then the contributions to the field are $\vec{e}_z I/(4a)(\varphi_0/\pi)$ and $-\vec{e}_z I/(4a)(1-\varphi_0/\pi)$, and in total the magnetic field in $(z=0;\, \varrho=0)$ would be

$$\vec{H} = \vec{e}_z I/(4a)(2\varphi_0/\pi - 1).$$

## 4.2  Magnetic Field of a Current Sheet

Consider a circular ring with outer radius $a$, inner radius $b > a$, conductivity $\kappa$ and a very small thickness $s \ll a$. The ring is open between $-\alpha < \varphi < \alpha$ and a constant current $I$ is impressed at its ends, which are perfectly conducting.

Calculate the magnetic field in the center of the ring.

The current distribution and the excited magnetic field can be determined independently in case of currents constant in time. As the thickness $s$ is small and the electrodes for the current supply are perfectly conducting, it is possible to describe the current distribution as a current sheet $\vec{K}(\varrho) = \vec{e}_\varphi K(\varrho)$. It is

$$\int\limits_1^2 \vec{E}\, d\vec{s} = -\int\limits_1^2 \operatorname{grad} V d\vec{s} = -\int\limits_1^2 dV = V_1 - V_2 =$$

$$= V(\varrho, \varphi = \alpha) - V(\varrho, \varphi = 2\pi - \alpha) = U = 2(\pi - \alpha)\,\varrho\, E(\varrho); \quad a < \varrho < b$$

$$K(\varrho) = \kappa s\, E(\varrho) = \frac{\kappa s U}{2(\pi-\alpha)\varrho}\,; \qquad \int\limits_{\varrho=a}^b K(\varrho)d\varrho = I = \frac{\kappa s U}{2(\pi-\alpha)}\ln(b/a)$$

$$K(\varrho) = \frac{I}{\varrho \ln(b/a)}\,.$$

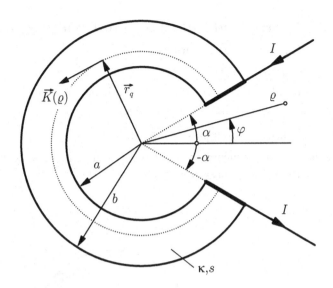

Now the magnetic field results from

$$\vec{H} = \frac{1}{4\pi} \int\limits_{a_K} \vec{K} \times \vec{r}/r^3 \, da \; ; \qquad \vec{r} = \vec{r}_p - \vec{r}_q,$$

where $a_K$ is the surface with the given current distribution.

For the calculation of the field in the center $\vec{r}_p = 0$ we first calculate the contribution of a differential current $K(\varrho) \, d\varrho$ on a circle of radius $\varrho$.

$$d\vec{H} = \frac{1}{4\pi} \int\limits_{\varphi=\alpha}^{2\pi-\alpha} \vec{e}_\varphi K(\varrho) \, d\varrho \times \frac{-\vec{e}_\varrho \varrho}{\varrho^3} \varrho \, d\varphi = \vec{e}_z \, \frac{\pi-\alpha}{2\pi} K(\varrho) \frac{d\varrho}{\varrho}$$

Finally a second integration leads to the magnetic field

$$\vec{H} = \vec{e}_z \frac{\pi-\alpha}{2\pi} \int\limits_{\varrho=a}^{b} K(\varrho) \frac{d\varrho}{\varrho} = \vec{e}_z \frac{\pi-\alpha}{2\pi} \frac{I}{\ln(b/a)} \int\limits_{\varrho=a}^{b} \frac{d\varrho}{\varrho^2}$$

$$\vec{H} = \vec{e}_z \frac{\pi-\alpha}{2\pi a} \frac{I}{\ln(b/a)} (1 - a/b) \, .$$

In the limit $a \to b$ the field becomes $\quad \vec{H} = \vec{e}_z (\pi - \alpha) \dfrac{I}{2\pi a} \, .$

## 4.3 Energy and Inductance of Conductors with Circular Symmetry

In this problem the energy of the magnetic field and the inductance of conductors with circular symmetry should be examined.

Consider a hollow cylinder of infinite length with radius $\varrho = a$ and permeability $\mu_0$ inside the homogeneous space of permeability $\mu$. The surface of the cylinder $\varrho = a$ carries the current sheet $\vec{K} = \vec{e}_z K_0$.

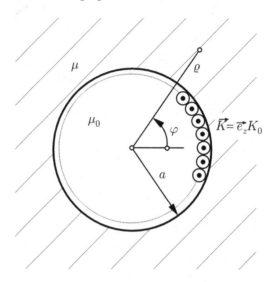

After a straightforward calculation the magnetic field becomes

$$\vec{H} = \vec{e}_\varphi \, K_0 \begin{cases} 0 & ; \quad \varrho < a \\ a/\varrho & ; \quad \varrho > a \end{cases}.$$

With this result the energy in the magnetic field per unit length $W_m/l$ would lead to the integral

$$W_m/l = \frac{1}{2} \int_a \vec{H}\vec{B}da = \frac{1}{2} \int\limits_{\varphi=0}^{2\pi} \int\limits_{\varrho=a}^{\infty} \mu K_0^2 \left(\frac{a}{\varrho}\right)^2 \varrho d\varrho d\varphi = \frac{\mu}{2} K_0^2 2\pi a^2 \lim_{c\to\infty} \int\limits_a^c \frac{d\varrho}{\varrho},$$

which is not convergent in the limit $c \to \infty$. Obviously the field is non-regular and thus the problem is not well-defined. (Because of the endless inductance it is not possible to impress the current.)

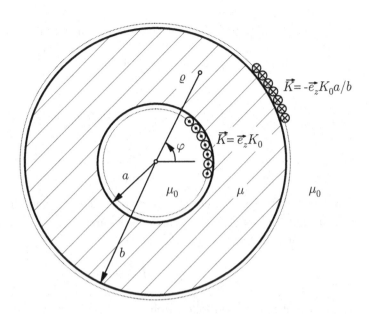

Now the problem is modified by adding a second current sheet $\vec{K} = -\vec{e}_z K_0 a/b$ at $\varrho = b > a$ and the permeability in $\varrho > b$ is $\mu_0$.

The solution for the magnetic field is

$$\vec{H} = \vec{e}_\varphi K_0 \begin{cases} 0 & ; \quad \varrho < a \\ a/\varrho & ; \quad a < \varrho < b \\ 0 & ; \quad \varrho > b \end{cases}$$

and the energy of the magnetic field per unit length $l$ becomes

$$W_m/l = \frac{1}{2}\mu \int\limits_{\varphi=0}^{2\pi} \int\limits_{\varrho=0}^{a} K_0^2 \left(\frac{a}{\varrho}\right)^2 \varrho d\varrho d\varphi = \mu K_0^2 a^2 \pi \ln \frac{b}{a} = \frac{1}{2} L/l \, I^2 \,.$$

The problem is regular and with the current $I = 2a\pi K_0$ the inductance per unit length $L/l$ is given by

$$L/l = \frac{\mu}{2\pi} \ln \frac{b}{a} \,.$$

This result is independent of $\mu_0$, because there is no magnetic field in the surrounding medium, i.e. the total current is zero.

It is possible to describe the magnetic field by means of a vector potential $\vec{A} = \vec{e}_z A(\varrho)$:

$$\vec{H} = \frac{1}{\mu} \text{rot} \, \vec{A} = -\vec{e}_\varphi \frac{1}{\mu} \frac{dA}{d\varrho} \,; \qquad A(\varrho) = -\mu K_0 \int\limits_{c}^{\varrho} \frac{a}{\varrho'} d\varrho' = -\mu K_0 a \ln \frac{\varrho}{c} \,.$$

In this case we can proof the calculation of the magnetic energy by the relation

$$W_m/l = \frac{1}{2} \int_a \vec{A}\vec{J} da = \frac{1}{2} \int_s \vec{A}\vec{K} ds = \frac{1}{2} L/l\, I^2$$

$$= \frac{1}{2} \int_0^{2\pi} [A(a)K(a)a - A(b)K(b)b]\, d\varphi = \frac{\mu}{4\pi}(2\pi a K_0)^2 \ln\frac{b}{a}$$

and get the same result as before. Thus if the total current in planes $z = $ const vanishes, then the problem is regular and the relation for the inductance is valid.

Now we modify the problem a second time. The current sheet on the inner surface vanishes and the outer surface carries the current sheet $\vec{K} = \vec{e}_z K_0 \cos\varphi$. Again the total current vanishes:

$$I = \int_0^{2\pi} K(\varrho) b\, d\varphi = 0\,.$$

Vector potential and field follow from

$$\vec{A} = \vec{e}_z A(\varrho,\varphi)\,; \quad \Delta A(\varrho,\varphi) = 0\,; \quad \varrho \neq b$$

$$\vec{H} = \frac{1}{\mu}\operatorname{rot}\vec{A} = \vec{e}_\varrho H_\varrho(\varrho,\varphi) + \vec{e}_\varphi H_\varphi(\varrho,\varphi) = \frac{1}{\mu}\left[\vec{e}_\varrho \frac{1}{\varrho}\frac{\partial A}{\partial\varphi} - \vec{e}_\varphi \frac{\partial A}{\partial\varrho}\right].$$

Here the general solution function of Laplace's equation for the vector potential is

$$A(\varrho,\varphi) = \sum_{n=1}^{\infty}\left[a_n\left(\frac{\varrho}{a}\right)^n + b_n\left(\frac{a}{\varrho}\right)^n\right][c_n\cos(n\varphi) + d_n\sin(n\varphi)]\,.$$

With the given current distribution the following approach applies

$$A(\varrho,\varphi) = \sum_{n=1}^{\infty}\left[\begin{array}{l} c_n(\varrho/a)^n \\ a_n(\varrho/a)^n + b_n(a/\varrho)^n \\ d_n(b/\varrho)^n \end{array}\right]\cos(n\varphi)\,; \quad \begin{array}{l} \varrho < a \\ a < \varrho < b \\ \varrho > b\,, \end{array}$$

where the constants $a_n, b_n, c_n$, and $d_n$ have been redefined. The solutions with $n = 0$ can be dropped as the total current vanishes.

For the determination of the constants firstly the continuity relation for the normal component of the magnetic flux density, which is in this case identical to the continuity of the vector potential, is evaluated with the result

$$c_n = a_n + b_n$$

$$d_n = a_n\left(\frac{b}{a}\right)^n + b_n\left(\frac{a}{b}\right)^n\,.$$

An additional equation results from the continuity of the tangential component of the magnetic field in $\varrho = a$:

$$H_\varphi\big|_{\substack{a<\varrho<b\\ \varrho\to a}} = H_\varphi\big|_{\substack{\varrho<a\\ \varrho\to a}} \; ; \qquad \frac{1}{\mu_0}\frac{\partial A}{\partial\varrho}\bigg|_{\substack{\varrho<a\\ \varrho\to a}} = \frac{1}{\mu}\frac{\partial A}{\partial\varrho}\bigg|_{\substack{a<\varrho<b\\ \varrho\to a}}$$

$$\Rightarrow \quad c_n = \frac{\mu_0}{\mu}[a_n - b_n] = a_n + b_n.$$

Finally on $\varrho = b$ applies

$$H_\varphi\big|_{\substack{\varrho>b\\ \varrho\to b}} - H_\varphi\big|_{\substack{a<\varrho<b\\ \varrho\to b}} = -\frac{1}{\mu_0}\frac{\partial A}{\partial\varrho}\bigg|_{\substack{\varrho>b\\ \varrho\to b}} + \frac{1}{\mu}\frac{\partial A}{\partial\varrho}\bigg|_{\substack{a<\varrho<b\\ \varrho\to b}} = K_0\cos\varphi$$

$$\sum_{n=1}^\infty \left[\frac{1}{\mu_0}\frac{n}{b}d_n + \frac{1}{\mu}\frac{n}{a}\left[a_n\left(\frac{b}{a}\right)^{n-1} - b_n\left(\frac{a}{b}\right)^{n+1}\right]\right]\cos(n\varphi) = K_0\cos\varphi$$

$$\Rightarrow \quad n = 1\,; \qquad d_1 + \frac{\mu_0}{\mu}\left[a_1\left(\frac{b}{a}\right) - b_1\left(\frac{a}{b}\right)\right] = \mu_0 K_0 b.$$

The solution of the system of equations is

$$\left[a_1\left(1+\frac{\mu_0}{\mu}\right)\frac{b}{a} + b_1\left(1-\frac{\mu_0}{\mu}\right)\frac{a}{b}\right] = \mu_0 K_0 b$$

$$a_1 = \frac{\mu_0\,K_0 b}{(1+\mu_0/\mu)b/a - k(1-\mu_0/\mu)a/b}\;; \qquad a_1\big|_{\mu\to\infty} = \frac{\mu_0\,K_0 a}{1-(a/b)^2}$$

$$b_1 = -a_1 k\,; \qquad k = \frac{\mu-\mu_0}{\mu+\mu_0}\;; \qquad b_1\big|_{\mu\to\infty} = \frac{-\mu_0\,K_0 a}{1-(a/b)^2}$$

$$c_1 = a_1(1-k)\,; \qquad c_1\big|_{\mu\to\infty} = 0\,; \qquad d_1 = a_1[b/a - k\,a/b]\,; \qquad d_1\big|_{\mu\to\infty} = \mu_0 K_0 b.$$

With this results the energy of the magnetic field calculates to

$$W_m/l = \frac{1}{2}\mu\int_{\varphi=0}^{2\pi}\int_{\varrho=0}^\infty H^2(\varrho,\varphi)\varrho\,d\varrho\,d\varphi = \frac{1}{2}\int_{\varphi=0}^{2\pi}\vec{K}(b)\,\vec{A}(b)\,b\,d\varphi =$$

$$= \frac{1}{2}bK_0 a_1\left[\frac{b}{a} - k\frac{a}{b}\right]\int_0^{2\pi}\cos^2\varphi\,d\varphi = \frac{\pi}{2}b\,K_0 a_1\left[\frac{b}{a} - k\frac{a}{b}\right].$$

The direct calculation of the integral of the square of the magnetic field over the cross section is more complicated. In the limit $\mu\to\infty$ the energy is

$$W_m/l\big|_{\mu\to\infty} = \frac{\pi}{2}\mu_0 K_0^2 b^2.$$

Although the total current vanishes in this problem, and thus it is a regular problem, it is not possible to define a reasonable inductance, because the current sheet is a continuous function of $\varphi$.

## 4.4  Shielding of the Magnetic Field of a Parallel-Wire Line

A parallel-wire line is placed inside a hollow cylinder of material with permeability $\mu$ and with radii $a$ and $b > a$. The two wires with currents $\pm I$ are at positions $(\varrho = c, \varphi = \pm\varphi_0)$ with $a > c$. The remaining space is of permeability $\mu_0$.

Calculate the magnetic field and give the limits $\mu = \mu_0, \mu \to \infty$, and $b \to \infty$. Use both the scalar potential and the vector potential.

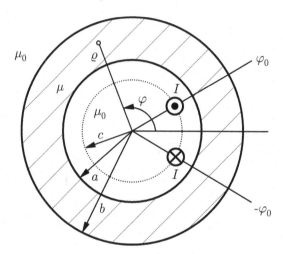

As the conductors of the excitation currents are thin wires, it is possible to apply a scalar magnetic potential $V_m$. The line currents at positions $(\varrho = c; \varphi = \pm\varphi_0)$ can be equivalently replaced by a magnetic dipole layer $\vec{m}_m = \vec{e}_\varrho \mu_0 I$ on the surface $(\varrho = c; -\varphi_0 < \varphi < \varphi_0)$. The field equations are

$$\text{rot}\,\vec{H} = 0; \quad \vec{B} = \mu\vec{H}; \quad \vec{H} = -\text{grad}\,V_m; \quad \Delta V_m = 0; \quad \varrho \neq c.$$

At first the exciting field $\vec{H}_e = -\text{grad}\,V_{me}$ of the line currents and thus that of the equivalent dipole layer in the homogeneous space of permeability $\mu_0$ is determined.

$$V_{me} = V_{me}(\varrho, \varphi); \quad \Delta V_{me} = \frac{1}{\varrho}\frac{\partial}{\partial\varrho}\left(\varrho\frac{\partial V_{me}}{\partial\varrho}\right) + \frac{1}{\varrho^2}\frac{\partial^2 V_{me}}{\partial\varphi^2} = 0$$

$$V_{me}(\varrho, \varphi) = R(\varrho)\Phi(\varphi)$$

$$\vec{H}_e = -\text{grad}\,V_{me} = -\vec{e}_\varrho\frac{\partial V_{me}}{\partial\varrho} - \vec{e}_\varphi\frac{1}{\varrho}\frac{\partial V_{me}}{\partial\varphi} = \vec{B}_e/\mu_0$$

The solution functions of the differential equation are powers of the axial distance $R_n(\varrho) = \{\varrho^n; \varrho^{-n}\}$ and the trigonometric functions $\Phi_n(\varphi) = \{\cos(n\varphi); \sin(n\varphi)\}$.

Due to the symmetry of the field with respect to $\varphi = 0$ and $\varphi = \pi$ the approach for the potential $V_{me}$ is given by

$$V_{me}(\varrho, \varphi) = \left\{ \begin{array}{l} E_{01} \\ E_{02} + D_{02} \ln \varrho / \ln c \end{array} \right\} + \sum_{n=1}^{\infty} \left\{ \begin{array}{l} E_{n1}(\varrho/c)^n \\ E_{n2}(c/\varrho)^n \end{array} \right\} \cos(n\varphi) ; \quad \begin{array}{l} \varrho < c \\ \varrho > c. \end{array}$$

The normal component $\vec{e}_\varrho \, B_{e\varrho}$ is continuous when passing the dipole layer at $\varrho = c$. This results in

$$\left. \frac{\partial V_{me}}{\partial \varrho} \right|_{\substack{\varrho > c \\ \varrho \to c}} = \left. \frac{\partial V_{me}}{\partial \varrho} \right|_{\substack{\varrho < c \\ \varrho \to c}} ; \qquad E_{n1} = -E_{n2} = -E_n ; \qquad D_{02} = 0 .$$

The potential itself is discontinuous at $(\varrho = c, |\varphi| < \varphi_0)$.

$$\left. V_{me} \right|_{\substack{\varrho > c \\ \varrho \to c}} - \left. V_{me} \right|_{\substack{\varrho < c \\ \varrho \to c}} = \left\{ \begin{array}{ll} m_m/\mu_0 = I & ; \quad |\varphi| < \varphi_0 \\ 0 & ; \quad \varphi_0 < \varphi < 2\pi - \varphi_0 \end{array} \right.$$

$$\sum_{n=0}^{\infty} (E_{n2} - E_{n1}) \cos(n\varphi) = \left\{ \begin{array}{ll} I & ; \quad |\varphi| < \varphi_0 \\ 0 & ; \quad \varphi_0 < \varphi < 2\pi - \varphi_0 \end{array} \right.$$

Now making use of the orthogonality relation for the trigonometric functions by multiplication with $\cos(k\varphi)$ and integration from $\varphi = 0$ to $\varphi = 2\pi$ leads to

$$2E_n\pi = I \int_{-\varphi_0}^{\varphi_0} \cos(n\varphi) d\varphi = I \frac{2}{n} \sin(n\varphi_0) ; \quad E_n = \frac{I}{n\pi} \sin(n\varphi_0) ; \quad E_0 = \frac{I\varphi_0}{\pi} .$$

The coefficient $E_0 = E_{02} - E_{01}$ in the Fourier series expansion of the potential difference leads to a source field. Hence this part does not contribute to the magnetic field and can be neglected.

Therewith the exciting potential is

$$V_{me} = \frac{I}{\pi} \sum_{n=1}^{\infty} \left\{ \begin{array}{l} -(\varrho/c)^n \\ (c/\varrho)^n \end{array} \right\} \frac{\sin(n\varphi_0)}{n} \cos(n\varphi).$$

A suitable approach for the resulting potential after the insertion of the permeable hollow cylinder is

$$V_m = \frac{I}{\pi} \sum_{n=1}^{\infty} \frac{\sin(n\varphi_0)}{n} \left\{ \begin{array}{l} \left\{ \begin{array}{l} -(\varrho/c)^n \\ (c/\varrho)^n \end{array} \right\} + A_n(\varrho/a)^n \\ B_n(\varrho/a)^n + C_n(a/\varrho)^n \\ D_n(b/\varrho)^n \end{array} \right\} \cos(n\varphi) ; \quad \begin{array}{l} \varrho < a \\ a < \varrho < b \\ \varrho > b . \end{array}$$

At the boundaries $\varrho = a$ and $\varrho = b$ the tangential component of the magnetic field is continuous. Therewith also the potential is continuous:

$$V_m\big|_{\substack{\varrho<a\\\varrho\to a}} = V_m\big|_{\substack{a<\varrho<b\\\varrho\to a}} ; \qquad V_m\big|_{\substack{a<\varrho<b\\\varrho\to b}} = V_m\big|_{\substack{\varrho>b\\\varrho\to b}}$$

$$(c/a)^n + A_n = B_n + C_n ; \qquad B_n(b/a)^n + C_n(a/b)^n = D_n .$$

Another two equations follow from the continuity of the normal component of the magnetic flux density:

$$\mu_0\,\frac{\partial V_m}{\partial \varrho}\bigg|_{\substack{\varrho<a\\\varrho\to a}} = \mu\frac{\partial V_m}{\partial \varrho}\bigg|_{\substack{a<\varrho<b\\\varrho\to a}} ; \qquad \mu\,\frac{\partial V_m}{\partial \varrho}\bigg|_{\substack{a<\varrho<b\\\varrho\to b}} = \mu_0\frac{\partial V_m}{\partial \varrho}\bigg|_{\substack{\varrho>b\\\varrho\to b}}$$

$$A_n - (c/a)^n = \mu/\mu_0(B_n - C_n); \qquad -D_n = \mu/\mu_0(B_n(b/a)^n - C_n(a/b)^n) .$$

Finally the solution of the four equations leads to the constants

$$A_n = 2\left[(c/a)^n + k(c/b)^n(a/b)^n\right]/M_n - (c/a)^n ; \qquad k = (\mu - \mu_0)/(\mu + \mu_0)$$

$$B_n = 2k(c/b)^n(a/b)^n/M_n ; \qquad C_n = 2(c/a)^n/M_n$$

$$D_n = 2(c/b)^n(1 + k)/M_n ; \qquad M_n = (1 + \mu/\mu_0)(1 - k^2(a/b)^{2n}) .$$

In the limit $\mu = \mu_0$ applies

$$k\big|_{\mu=\mu_0} = 0; \qquad M_n\big|_{\mu=\mu_0} = 2; \qquad A_n\big|_{\mu=\mu_0} = 0; \qquad B_n\big|_{\mu=\mu_0} = 0$$

$$C_n\big|_{\mu=\mu_0} = (c/a)^n ; \qquad D_n\big|_{\mu=\mu_0} = (c/b)^n$$

and the potential $V_m$ is equal to the exciting potential $V_{me}$.

The limit $\mu \to \infty$ results with $k \to 1$ and $M_n \to \infty$ in

$$A_n\big|_{\mu\to\infty} = -(c/a)^n ; \qquad B_n\big|_{\mu\to\infty} = C_n\big|_{\mu\to\infty} = D_n\big|_{\mu\to\infty} = 0 .$$

Now the resulting potential is

$$V_m\big|_{\mu\to\infty} = \frac{I}{\pi}\sum_{n=1}^{\infty}\frac{\sin(n\varphi_0)}{n}\left\{\left\{\begin{array}{c}-(\varrho/c)^n\\(c/\varrho)^n\end{array}\right\} - (\varrho/a)^n(c/a)^n\right\}\cos(n\varphi); \qquad \begin{array}{c}\varrho<c\\\varrho>c.\end{array}$$

This expression can be described only by means of the exciting potential

$$V_m\big|_{\mu\to\infty} = V_{me}(\varrho, \varphi) - V_{me}(a^2/\varrho, \varphi)$$

and thus we get the method of images for a highly permeable cylinder.

In case of a large radius $b \gg a$ we get with $b \to \infty$ the results

$$M_n|_{b\to\infty} = (1 + \mu/\mu_0) ; \qquad A_n|_{b\to\infty} = (c/a)^n \frac{1 - \mu/\mu_0}{1 + \mu/\mu_0} = -k(c/a)^n$$

$$B_n|_{b\to\infty} = 0 ; \qquad C_n|_{b\to\infty} = (c/a)^n \frac{2}{1 + \mu/\mu_0} .$$

The potential becomes

$$V_m|_{b\to\infty} = \frac{I}{\pi} \sum_{n=1}^{\infty} \frac{\sin(n\varphi_0)}{n} \left\{ \begin{array}{l} \left\{ \begin{array}{l} -(\varrho/c)^n \\ (c/\varrho)^n \end{array} \right\} - k(c/a)^n(\varrho/a)^n \\ 2/(1 + \mu/\mu_0)\,(c/\varrho)^n \end{array} \right\} \cos(n\varphi) ; \quad \begin{array}{l} \varrho < c \\ c < \varrho < a \\ \varrho > a . \end{array}$$

Again we can write the expression in terms of the exciting potential

$$V_m|_{b\to\infty} = \left\{ \begin{array}{ll} V_{me}(\varrho,\varphi) - kV_{me}(a^2/\varrho,\varphi) ; & \varrho < a \\ (1 - k)V_{me}(\varrho,\varphi) & ; \quad \varrho > a . \end{array} \right.$$

This result corresponds to the method of images for permeable cylinders.

It is also possible to solve this problem by means of a vector potential, which has the same direction as the line currents. Thus again the field is described by a scalar function, that is a solution of Laplace's equation.

$$\vec{A}(\varrho,\varphi) = \vec{e}_z A(\varrho,\varphi) ; \qquad \Delta A(\varrho,\varphi) = 0 ; \qquad \vec{B} = \mathrm{rot}\,\vec{A} = \vec{e}_\varrho \frac{1}{\varrho} \frac{\partial A}{\partial \varphi} - \vec{e}_\varphi \frac{\partial A}{\partial \varrho}$$

For the calculation of the exciting potential $\vec{A}_e(\varrho,\varphi) = \vec{e}_z A_e(\varrho,\varphi)$ in the homogeneous space of permittivity $\mu_0$ we replace the currents equivalently by a $z$-directed current sheet $K(\varphi)$ that vanishes except for $\varphi = \pm\varphi_0$. With respect to the symmetry the solutions of Laplace's equation $\Delta A_e(\varrho,\varphi) = 0$ are

$$A_e(\varrho,\varphi) = \sum_{n=1}^{\infty} A_n \left\{ \begin{array}{l} (\varrho/c)^n \\ (c/\varrho)^n \end{array} \right\} \sin(n\varphi) ; \qquad \begin{array}{l} \varrho \le c \\ \varrho \ge c . \end{array}$$

The solution with $n = 0$ is omitted because the total current vanishes.

In the following all constants have a different meaning than in the scalar potential approach.

The approach already satisfies the continuity of the normal component of the flux density $B_{e\varrho}$ in $\varrho = c$, which is identical to the continuity of the vector potential. For the tangential component $\vec{e}_\varphi \vec{B}_e/\mu$ it follows

$$H_{e\varphi}\Big|_{\substack{\varrho>c \\ \varrho\to c}} - H_{e\varphi}\Big|_{\substack{\varrho<c \\ \varrho\to c}} = K(\varphi) ; \qquad \frac{\partial A_e}{\partial \varrho}\bigg|_{\substack{\varrho<c \\ \varrho\to c}} - \frac{\partial A_e}{\partial \varrho}\bigg|_{\substack{\varrho>c \\ \varrho\to c}} = \mu_0 K(\varphi)$$

$$\sum_{n=1}^{\infty} 2n/c\, A_n \sin(n\varphi) = \mu_0 K(\varphi).$$

Making use of the orthogonality relation for trigonometric functions leads to

$$2n/c A_n \pi = \mu_0 \int_0^{2\pi} K(\varphi)\sin(n\varphi)d\varphi = 2\mu_0/c\sin(n\varphi_0)\underbrace{\int_0^{\pi} K(\varphi)c d\varphi}_{I} = \frac{2\mu_0 I}{c}\sin(n\varphi_0)$$

$$\Rightarrow \quad A_n = \frac{\mu_0 I}{\pi}\frac{\sin(n\varphi_0)}{n}.$$

Now the ansatz for the resulting potential after inserting the permeable cylinder is

$$A(\varrho,\varphi) = \frac{\mu_0 I}{\pi}\sum_{n=1}^{\infty}\frac{\sin(n\varphi_0)}{n}\left\{\begin{array}{ll}\left\{\begin{array}{l}(\varrho/c)^n\\(c/\varrho)^n\end{array}\right\} + B_n(\varrho/a)^n & \varrho < a\\ C_n(\varrho/a)^n + D_n(a/\varrho)^n & a < \varrho < b\\ E_n(b/\varrho)^n & \varrho > b.\end{array}\right\}\sin(n\varphi)\,;$$

The continuity of the normal component of the flux density and accordingly of the vector potential on the surfaces $\varrho = a$ and $\varrho = b$ yields

$$(c/a)^n + B_n = C_n + D_n\,; \qquad C_n(b/a)^n + D_n(a/b)^n = E_n$$

and from the continuity of the tangential component of the magnetic field it follows

$$-E_n = \mu_0/\mu\left[C_n(b/a)^n - D_n(a/b)^n\right]\,; \qquad -(c/a)^n + B_n = \mu_0/\mu\left[C_n - D_n\right].$$

Finally after solving the system of equations the constants are

$$B_n = 2(c/a)^n\left[1 - k(a/b)^{2n}\right]/M - (c/a)^n\,; \qquad E_n = 2(c/b)^n(1-k)/M$$

$$C_n = -2k(c/b)^n(a/b)^n/M; \qquad D_n = 2(c/a)^n/M$$

$$M = (1+\mu_0/\mu)(1 - k^2(a/b)^{2n})\,; \qquad k = (\mu - \mu_0)/(\mu + \mu_0).$$

In the limit $\mu = \mu_0$ we get with $k = 0$ and $M = 2$ the result

$$B_n|_{\mu=\mu_0} = 0\,; \quad C_n|_{\mu=\mu_0} = 0\,; \quad D_n|_{\mu=\mu_0} = (c/a)^n\,; \quad E_n|_{\mu=\mu_0} = (c/b)^n,$$

so that only the exciting potential remains.

With $b \to \infty$ the solution becomes

$$M_n|_{b\to\infty} = 1 + \frac{\mu_0}{\mu}\,; \quad B_n|_{b\to\infty} = k\left(\frac{c}{a}\right)^n\,; \quad C_n|_{b\to\infty} = 0\,; \quad D_n|_{b\to\infty} = \frac{2(c/a)^n}{1+\mu_0/\mu}$$

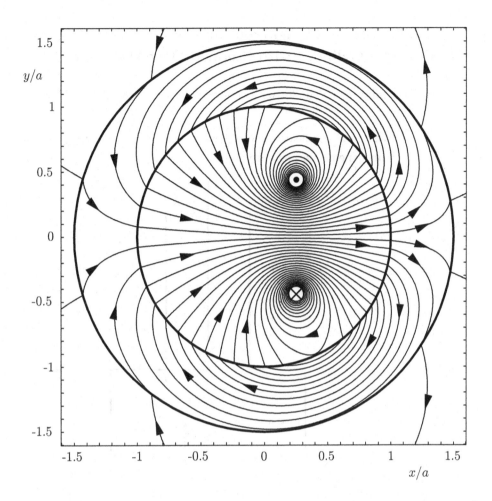

Fig. 4.4–1: Magnetic lines of force for $\mu/\mu_0 = 10$, $b/a = 1.5$, $c/a = 0.5$, and $\varphi_0 = \pi/3$

$$A(\varrho,\varphi)|_{b\to\infty} = \frac{\mu_0 I}{\pi} \sum_{n=1}^{\infty} \frac{\sin(n\varphi_0)}{n} \left\{ \begin{Bmatrix} (\varrho/c)^n \\ (c/\varrho)^n \end{Bmatrix} + k(c/a)^n(\varrho/a)^n \\ \dfrac{2}{1+\mu_0/\mu}\,(c/\varrho)^n \right\} \sin(n\varphi)$$

for $\varrho < a$ and $\varrho > a$. This expression has the form

$$A(\varrho,\varphi)|_{b\to\infty} = \begin{cases} A_e(\varrho,\varphi) + k\,A_e(a^2/\varrho,\varphi) & ; \quad \varrho < a \\ (1+k)A_e(\varrho,\varphi) & ; \quad \varrho > a. \end{cases}$$

It is the analog expression for vector potentials of the method of images for a permeable cylinder.

## 4.5  Magnetic Field and Stationary Current Flow in a Cylinder with Stepped Down Diameter

The radius of a cylinder with conductivity $\kappa$ is stepped down from $b$ in $z < 0$ to $a$ in $z > 0$.

Find the magnetic field inside the cylinder, when the total current $I_0$ is constant and the homogeneous permeability is $\mu$.

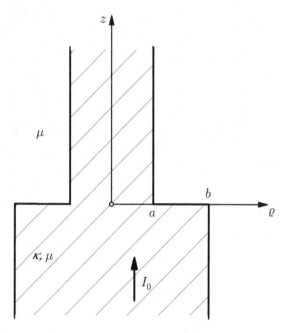

Electric and magnetic fields can be determined independently if all fields are constant in time. The current distribution of the present problem has been analyzed in chapter 3.

Outside the conducting material the $\varphi$-directed magnetic field is equal to the field of a line current $I_0$ on the $z$-axis. Inside the cylinder the magnetic field satisfies the differential equation

$$\text{rot rot } \vec{H} = \text{rot rot } [\vec{e}_\varphi\, H(\varrho, z)] = \vec{e}_\varphi \left[ -\frac{\partial^2 H}{\partial z^2} - \frac{\partial}{\partial \varrho} \left( \frac{1}{\varrho} \frac{\partial}{\partial \varrho} (\varrho\, H) \right) \right] = 0$$

$$\Rightarrow \quad \frac{\partial^2 H}{\partial \varrho^2} + \frac{1}{\varrho} \frac{\partial H}{\partial \varrho} - \frac{H}{\varrho^2} + \frac{\partial^2 H}{\partial z^2} = 0; \quad H(\varrho, z) = R(\varrho)\, Z(z).$$

The solutions of this differential equation are the ordinary Bessel functions $R_m(\varrho) = \{J_1(m\varrho);\ N_1(m\varrho)\}$ and the exponential functions $Z_m(z) = \exp(\pm mz)$ with $m \neq 0$. In case of $m = 0$ the solution functions are powers of $\varrho$.

With $H_0 = I_0/(2\pi a)$ the ansatz for the magnetic field is

$$H_1(\varrho, z) = H_0 \left[ \frac{\varrho}{a} + \sum_{r=1}^{\infty} A_r J_1(x_{1r}\varrho/a) \exp(-x_{1r}z/a) \right] ; \quad \varrho \leq a , z \geq 0$$

$$H_2(\varrho, z) = H_0 \frac{a}{b} \left[ \frac{\varrho}{b} + \sum_{r=1}^{\infty} B_r J_1(x_{1r}\varrho/b) \exp(x_{1r}z/b) \right] ; \quad \varrho \leq b; z \leq 0.$$

The $z$-independent part describes the field for points $|z| \gg b$.

The unknown constants are the solution of a linear system of equations that results from the evaluation of the boundary conditions in $z = 0$. From the continuity of the magnetic field we obtain

$$\frac{a}{b} \left[ \frac{\varrho}{b} + \sum_{r=1}^{\infty} B_r J_1(x_{1r}\,\varrho/b) \right] = \begin{cases} \varrho/a + \sum_{r=1}^{\infty} A_r J_1(x_{1r}\varrho/a) ; & \varrho \leq a \\ a/\varrho & ; \quad a \leq \varrho \leq b. \end{cases}$$

Making use of the orthogonality relation for Bessel functions results after multiplication with $\varrho J_1(x_{1s}\,\varrho/b)$ and integration from $\varrho = 0$ to $\varrho = b$ in

$$\frac{a}{b} \left[ S_4 + B_s \frac{1}{2} b^2 J_0^2(x_{1s}) \right] = S_1 + \sum_{r=1}^{\infty} A_r S_2 + S_3$$

with

$$S_1 = \int_0^a \frac{\varrho}{a} J_1(x_{1s}\,\varrho/b)\, \varrho d\varrho = \frac{ab}{x_{1s}} J_2(x_{1s}\,a/b)$$

$$S_2 = \int_0^a J_1(x_{1r}\,\varrho/a) J_1(x_{1s}\,\varrho/b)\, \varrho d\varrho = \frac{x_{1r} J_1(x_{1s}\,a/b) J_0(x_{1r})}{(x_{1s}/b)^2 - (x_{1r}/a)^2}$$

$$S_3 = \int_a^b \frac{a}{\varrho} J_1(x_{1s}\,\varrho/b)\, \varrho d\varrho = \frac{ab}{x_{1s}} [J_0(x_{1s}\,a/b) - J_0(x_{1s})]$$

$$S_4 = \int_0^b \frac{\varrho}{b} J_1(x_{1s}\,\varrho/b)\, \varrho d\varrho = \frac{b^2}{x_{1s}} J_2(x_{1s}) .$$

The result is a linear system of equations

$$\frac{1}{2} \frac{b}{a} J_0^2(x_{1s}) B_s = 2 \frac{J_1(x_{1s}\,a/b)}{(x_{1s}\,a/b)^2} + J_1(x_{1s}\,a/b) \sum_{r=1}^{\infty} A_r \frac{x_{1r} J_0(x_{1r})}{(x_{1s}\,a/b)^2 - x_{1r}^2} ; \quad s = 1, 2, 3, \ldots .$$

Because of the homogeneous conductivity the tangential component of the current density $J_\varrho$ is continuous in $z = 0$ and $\varrho < a$:

$$J_\varrho|_{z>0,z\to 0} = J_\varrho|_{z<0,z\to 0} \quad \Rightarrow \quad \left.\frac{\partial H_1}{\partial z}\right|_{z=0} = \left.\frac{\partial H_2}{\partial z}\right|_{z=0} \quad \Rightarrow$$

$$-\sum_{r=1}^{\infty} A_r \frac{x_{1r}}{a} J_1(x_{1r}\varrho/a) = \frac{a}{b}\sum_{r=1}^{\infty} B_r \frac{x_{1r}}{b} J_1(x_{1r}\varrho/b) ; \qquad \varrho < a .$$

Now multiplication with $\varrho J_1(x_{1s}\varrho/a)$ and integration from $\varrho = 0$ to $\varrho = a$ leads to

$$A_s = \frac{2}{J_0(x_{1s})} \frac{a}{b} \sum_{r=1}^{\infty} x_{1r}\, a/b\, B_r \frac{J_1(x_{1r}a/b)}{x_{1s}^2 - (x_{1r}a/b)^2} ,$$

where the previously defined integral values have been used again.

Therewith all constants are defined by a system of equations and hence the magnetic field is known.

The current density is calculated by

$$\vec{J} = \mathrm{rot}\,\vec{H} = -\vec{e}_\varrho \frac{\partial H}{\partial z} + \vec{e}_z \frac{1}{\varrho}\frac{\partial}{\partial\varrho}(\varrho H)$$

and finally

$$\vec{J_1}(\varrho, z) = H_0/a \left[2\vec{e}_z + \sum_{r=1}^{\infty} x_{1r}\, A_r \left(\vec{e}_z\, J_0(x_{1r}\varrho/a) + \vec{e}_\varrho\, J_1\left(x_{1r}\varrho/a\right)\right) \exp(-x_{1r}z/a)\right]$$

$$\vec{J_2}(\varrho, z) = H_0 a/b^2 \left[2\vec{e}_z - \sum_{r=1}^{\infty} x_{1r}\, B_r \left(\vec{e}_\varrho\, J_1(x_{1r}\varrho/b) - \vec{e}_z\, J_0(x_{1r}\varrho/b)\right) \exp(x_{1r}z/b)\right].$$

For $z \to \pm\infty$ the current densities are homogeneous:

$$\vec{J_1} = 2H_0/a = I_0/(\pi a^2) \qquad \text{and} \qquad \vec{J_2} = 2H_0 a/b^2 = I_0/(\pi b^2).$$

## 4.6    Force on a Conductor Loop in Front of a Permeable Sphere

Consider a circular conductor loop at position $(r = b; \vartheta = \vartheta_0)$ in front of a permeable sphere of radius $a < b$ and permeability $\mu$. The remaining space is of permittivity $\mu_0$ and the loop carries the current $I$.

What is the force acting on the conductor loop? Analyze the limit $\mu \to \infty$. Finally derive the result for the field of a permeable sphere, when the exciting field is homogeneous.

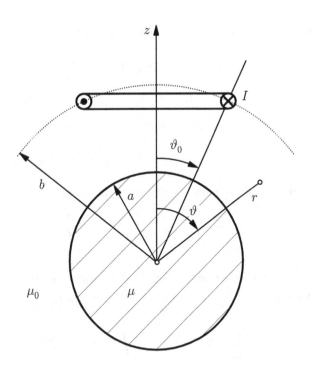

It is practical to use the scalar magnetic potential $V_m$ to calculate the magnetic field $\vec{H} = -\text{grad}\, V_m$. At first the exciting potential $V_{me}$ of the conductor loop in the homogeneous space of permeability $\mu_0$ is determined. For this purpose we consider the magnetic dipole layer $\vec{m}_m = \vec{r}/r\mu_0 I = \vec{r}/r\, m_m$ on the spherical cap $(r = b, \vartheta < \vartheta_0)$, that replaces the conductor loop equivalently.

The scalar magnetic potential

$$V_{me} = \frac{1}{4\pi\mu_0} \int_a \vec{m}_m \frac{\vec{r}}{r^3}\, da$$

must satisfy the boundary condition

$$\left[ V_{me}\Big|_{\substack{r>b \\ r\to b}} - V_{me}\Big|_{\substack{r<b \\ r\to b}} \right]_{0 \le \vartheta \le \vartheta_0} = \frac{m_m}{\mu_0} = I$$

on $r = b$. It is a solution of Laplace's equation $\Delta V_{me} = 0$, thus:

$$V_{me} = \sum_{n=0}^{\infty} \left\{ \begin{array}{c} A_n (b/r)^{n+1} \\ B_n (r/b)^n \end{array} \right\} P_n(\cos\vartheta); \qquad \begin{array}{c} r > b \\ r < b \end{array}$$

$$\vec{H}_e = -\text{grad}\, V_{me} = -\vec{e}_r \frac{\partial V_{me}}{\partial r} - \vec{e}_\vartheta \frac{1}{r} \frac{\partial V_{me}}{\partial \vartheta}\,.$$

The continuity of the normal component of the magnetic flux density

$$\vec{r}/r \left( \vec{B} \Big|_{\substack{r>b \\ r \to b}} - \vec{B} \Big|_{\substack{r<b \\ r \to b}} \right) = 0$$

gives

$$\mu_0 \frac{\partial V_{me}}{\partial r} \Big|_{\substack{r>b \\ r \to b}} = \mu_0 \frac{\partial V_{me}}{\partial r} \Big|_{\substack{r<b \\ r \to b}}$$

$$\Rightarrow \quad -(n+1) A_n = n B_n; \qquad A_n = -\frac{n}{n+1} B_n.$$

And from the boundary condition for the potential at the dipole layer we get

$$\sum_{n=0}^{\infty} (A_n - B_n) P_n(\cos \vartheta) = \begin{cases} I & ; \quad 0 \le \vartheta < \vartheta_0 \\ 0 & ; \quad \vartheta_0 < \vartheta \le \pi \end{cases}.$$

The unknown constants are derived by means of the orthogonality of the Legendre polynomials $P_n$. With $u = \cos \vartheta$ and $u_0 = \cos \vartheta_0$ we get

$$\sum_{n=0}^{\infty} (A_n - B_n) \cdot \underbrace{\int_{-1}^{+1} P_n(u) P_k(u) du}_{} = I \int_{u_0}^{1} P_k(u) du$$

$$= \begin{cases} 2/(2k+1) & ; \quad n = k \\ 0 & ; \quad n \ne k \end{cases}$$

$$(A_n - B_n) \frac{2}{2n+1} = I \left. \frac{P_{n+1} - P_{n-1}}{2n+1} \right|_{u_0}^{1}; \qquad P_n(1) = 1$$

$$A_n - B_n = \frac{I}{2} [P_{n-1}(u_0) - P_{n+1}(u_0)] = -\frac{2n+1}{n+1} B_n = \frac{2n+1}{n} A_n$$

$$K_n = \frac{P_{n-1}(u_0) - P_{n+1}(u_0)}{2n+1}; \qquad A_n = n \frac{I}{2} K_n; \qquad B_n = -(n+1) \frac{I}{2} K_n.$$

It should be noted, that the complementary magnetic dipole layer $-\vec{r}/r\mu_0 I$ on the spherical segment $(r = b; \vartheta_0 < \vartheta \le \pi)$ would lead to the same result.

After insertion of the permeable sphere the approach for the resulting potential $V_m$ with $\Delta V_m = 0$ becomes

$$V_m = \begin{Bmatrix} V_{me} \\ 0 \end{Bmatrix} + \sum_{n=0}^{\infty} \begin{Bmatrix} C_n (a/r)^{n+1} \\ D_n (r/a)^n \end{Bmatrix} P_n(\cos \vartheta); \qquad \begin{array}{l} r > a \\ r < a \end{array}.$$

For points $r < a$ the exciting potential has the same dependencies on the coordinates $\varrho$ and $\vartheta$ as the added potential, thus it is omitted here.

The tangential component of the magnetic field and accordingly the potential is continuous in $r = a$:

$$-(n+1)\frac{I}{2} K_n \left(\frac{a}{b}\right)^n + C_n = D_n \,.$$

A second condition is the continuity of the magnetic flux density in $r = a$:

$$\mu_0 \left[ B_n \left(\frac{a}{b}\right)^{n-1} n\frac{1}{b} - (n+1) C_n \frac{1}{a} \right] = \mu \frac{n}{a} D_n \,.$$

Hence the solution is

$$C_n = \frac{n(n+1)(1-\mu_0/\mu)}{n+(n+1)\mu_0/\mu} \frac{I}{2} K_n \left(\frac{a}{b}\right)^n$$

$$D_n = -\frac{(n+1)(2n+1)\mu_0/\mu}{n+(n+1)\mu_0/\mu} \frac{I}{2} K_n \left(\frac{a}{b}\right)^n \,.$$

In the limit $\mu \to \mu_0$ again only the exciting potential remains. The limit $\mu \to \infty$ leads to

$$C_n|_{\mu\to\infty} = (n+1)\frac{I}{2} K_n \left(\frac{a}{b}\right)^n \,; \qquad D_n|_{\mu\to\infty} = 0$$

$$V_m|_{\mu\to\infty} = V_{me} + \frac{I}{2} \sum_{n=1}^{\infty} (n+1) K_n \left(\frac{a}{b}\right)^n \left(\frac{a}{r}\right)^{n+1} P_n(\cos\vartheta) \,; \qquad r \geq a$$

$$= V_{me}(r,\vartheta) - \left(\frac{a}{r}\right) V_{me} \left(a^2/r, \vartheta\right) \,; \qquad r \geq a \,.$$

This relation is the method of images for magnetic fields at highly permeable spheres. Unlike the method of images for permeable half-spaces and permeable cylinders, in case of the sphere we can write this law only for $\mu \to \infty$.

For $\vartheta_0 = \pi/2$, $b \gg a$, and $I/b = \mathrm{const}$ the exciting magnetic field acting on the sphere is almost homogeneous and $z$-directed

$$\vec{H}_e = \vec{e}_z H_0 = \vec{e}_z I/(2b) = (\vec{e}_r \cos\vartheta - \vec{e}_\vartheta \sin\vartheta) H_0 =$$

$$= -\mathrm{grad}\ V_{me}|_{b\to\infty} = \mathrm{grad}\left(H_0 r \cos\vartheta\right) \,.$$

Thus in the approach for the resulting potential we just need terms with $n = 1$.

Hence the resulting potential is

$$V_m|_{b\to\infty} = \begin{cases} B_1(r/b) + C_1(a/r)^2 \\ D_1(r/a) \end{cases} P_1(\cos\vartheta)$$

$$= -H_0 a \begin{cases} r/a - \dfrac{1-\mu_0/\mu}{1+2\mu_0/\mu}(a/r)^2 \\[2mm] \dfrac{3\mu_0/\mu}{1+2\mu_0/\mu}(r/a) \end{cases} \cos\vartheta\,; \qquad \begin{matrix} r \geq a \\[4mm] r \leq a\,. \end{matrix}$$

The force acting on the conductor loop with contour $C$ is $z$-directed.

$$\vec{F} = I \oint_C d\vec{s} \times \vec{B}_s = \vec{e}_z F\,; \qquad F = I \oint_C \vec{e}_z(d\vec{s} \times \vec{B}_s) = -I \oint_C (\vec{e}_z \times \vec{B}_s)d\vec{s}$$

$$\vec{e}_z = \vec{e}_r \cos\vartheta - \vec{e}_\vartheta \sin\vartheta\,; \qquad \vec{B}_s = \vec{e}_r B_{sr} + \vec{e}_\vartheta B_{s\vartheta}\,; \qquad d\vec{s} = b\sin\vartheta_0\,\vec{e}_\varphi\,d\varphi$$

This results in

$$F = -I\,b\sin\vartheta_0\,2\pi\,[\cos\vartheta_0\,B_{s\vartheta}(b,\vartheta_0) + \sin\vartheta_0\,B_{sr}(b,\vartheta_0)]$$

$$\vec{B}_s = -\mu_0 \mathrm{grad}\,(V_m - V_{me})$$

$$B_{sr}(b,\vartheta_0) = -\mu_0 \frac{\partial}{\partial r}\,(V_m - V_{me})\Big|_{\substack{r=b\\ \vartheta=\vartheta_0}} = \mu_0 \sum_{n=0}^{\infty} C_n \frac{(n+1)}{a} \left(\frac{a}{b}\right)^{n+2} P_n(\cos\vartheta_0)$$

$$B_{s\vartheta}(b,\vartheta_0) = -\frac{\mu_0}{r}\frac{\partial}{\partial\vartheta}\,(V_m - V_{me})\Big|_{\substack{r=b\\ \vartheta=\vartheta_0}} = \frac{\mu_0}{b} \sum_{n=0}^{\infty} C_n \left(\frac{a}{b}\right)^{n+1} P_n'(\cos\vartheta_0)\sin\vartheta_0$$

$$F = -2\pi\,\mu_0 I \sin^2\vartheta_0 \sum_{n=0}^{\infty} C_n \left(\frac{a}{b}\right)^{n+1} [(n+1)P_n(\cos\vartheta_0) + \cos\vartheta_0\,P_n'(\cos\vartheta_0)]\,.$$

## 4.7 Shielding of a Homogeneous Magnetic Field by a Permeable Hollow Cylinder

Consider a permeable hollow cylinder of permeability $\mu$ and with radii $\varrho = a$ and $\varrho = b > a$. The remaining space is of permeability $\mu_0$. An homogeneous exciting field $\vec{H}_e = \vec{e}_x H_e$ is acting in perpendicular direction to the cylinder axis.

Calculate the resulting magnetic field $\vec{H}$ and the screening factor $\eta_s = |\vec{H}_e|/|\vec{H}|$ for points inside the cylinder $\varrho < a$. In addition, give the limits $\mu = \mu_0$ and $\mu \to \infty$.

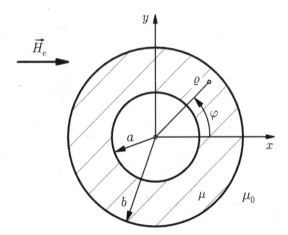

The homogeneous field might emerge from two homogeneous current sheets in planes $y = \text{const}$ with current flow in opposite $z$-direction. Thus it is possible to describe the magnetic field by a $z$-directed vector potential $\vec{A} = \vec{e}_z A(\varrho, \varphi)$.

From the vector potential $\vec{A}_e = \vec{e}_z A_e(\varrho, \varphi)$ of the exciting field the magnetic flux density $\vec{B}_e = \text{rot}\,\vec{A}_e = \mu_0 \vec{H}_e = \vec{e}_x \mu_0 H_e$ is calculated to

$$\vec{B}_e = \text{grad}\,A_e \times \vec{e}_z = \left[ \vec{e}_\varrho \frac{\partial A_e}{\partial \varrho} + \frac{\vec{e}_\varphi}{\varrho} \frac{\partial A_e}{\partial \varphi} \right] \times \vec{e}_z = \frac{\vec{e}_\varrho}{\varrho} \frac{\partial A_e}{\partial \varphi} - \vec{e}_\varphi \frac{\partial A_e}{\partial \varrho}$$

$$= \mu_0 \vec{H}_e = \mu_0 H_e \vec{e}_x = \mu_0 H_e \left[ \vec{e}_\varrho \cos\varphi - \vec{e}_\varphi \sin\varphi \right]$$

$$\frac{1}{\varrho} \frac{\partial A_e}{\partial \varphi} = \mu_0 H_e \cos\varphi \,; \qquad \frac{\partial A_e}{\partial \varrho} = \mu_0 H_e \sin\varphi \,.$$

The vector potential satisfies the differential equation $\Delta A_e(\varrho, \varphi) = 0$ and the general solution is

$$A_e(\varrho, \varphi) = (a_0 + b_0 \ln\varrho)(c_0 + d_0\varphi) + \sum_{n=1}^{\infty} (a_n \varrho^n + b_n \varrho^{-n})(c_n \cos(n\varphi) + d_n \sin(n\varphi)) \,.$$

Because of the given exciting field only the solutions with $n = 1$ are needed and furthermore $b_1 = 0$, $c_1 = 0$, and $d_1 = 1$.

$$\Rightarrow \quad A_e(\varrho, \varphi) = a_1 \varrho \sin\varphi = \mu_0 H_e \varrho \sin\varphi$$

Accordingly the ansatz for the resulting vector potential $A(\varrho, \varphi)$ in presence of the cylinder is

$$A(\varrho, \varphi) = \mu_0 H_e b \left\{ \begin{array}{l} a_1 \varrho/a \\ a_2 \varrho/a + b_2 a/\varrho \\ \varrho/b + b_3 b/\varrho \end{array} \right\} \sin(\varphi) \,; \qquad \begin{array}{l} \varrho \le a \\ a \le \varrho \le b \\ \varrho \ge b. \end{array}$$

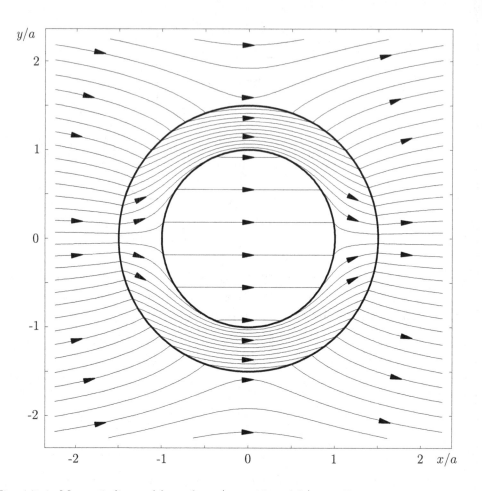

Fig. 4.7–1: Magnetic lines of force for $\mu/\mu_0 = 10$ and $b/a = 1.5$

The evaluation of the boundary conditions leads to the unknown constants. The continuity of the normal component of the magnetic flux density is identical to the continuity of the vector potential:

$$a_1 = a_2 + b_2; \qquad a_2 + b_2 \left(\frac{a}{b}\right)^2 = \frac{a}{b}(1 + b_3).$$

Another two equations follow from the continuity of the tangential component $H_\varphi$ of the magnetic field:

$$a_1 = \frac{\mu_0}{\mu}(a_2 - b_2); \qquad a_2 - b_2 \left(\frac{a}{b}\right)^2 = \frac{\mu}{\mu_0}\frac{a}{b}(1 - b_3).$$

Now there are four equations for four unknowns. With $k = (\mu - \mu_0)/(\mu + \mu_0)$ and

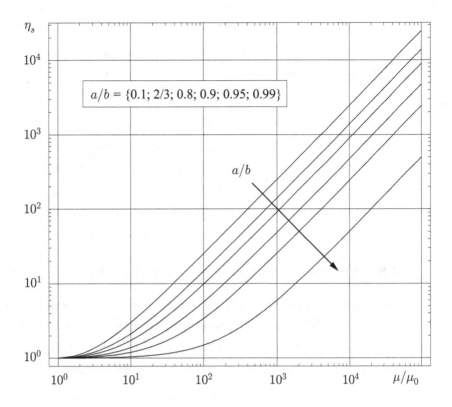

Fig. 4.7–2: Screening factor in dependence on the permeability

$M = (1 + \mu_0/\mu)(1 - k^2(a/b)^2)$ the solutions are

$$b_2 = -\frac{2\,(a/b)\,k}{M}\,; \qquad a_2 = \frac{2\,a/b}{M}\,; \qquad a_1 = 2\frac{a}{b}\frac{1-k}{M}\,; \qquad b_3 = \frac{2(1-k(a/b)^2)}{M} - 1\,.$$

In the limit $\mu \to \mu_0$ the results are

$$a_1\big|_{\mu\to\mu_0} = a/b\,; \qquad a_2\big|_{\mu\to\mu_0} = a/b\,; \qquad b_2\big|_{\mu\to\mu_0} = 0\,; \qquad b_3\big|_{\mu\to\mu_0} = 0$$

and it remains only the exciting potential.

The other case of a very high permeability results in the limit $\mu \to \infty$ in

$$a_1\big|_{\mu\to\infty} = 0\,; \qquad b_2\big|_{\mu\to\infty} = \frac{-2a/b}{1 - (a/b)^2}\,; \qquad a_2\big|_{\mu\to\infty} = -b_2\big|_{\mu\to\infty}\,; \qquad b_3\big|_{\mu\to\infty} = 1\,.$$

The magnetic field vanishes inside the cylinder $\varrho < b$ and on the cylinder $\varrho = b$ the field has only a radial component $H_\varrho$.

With a finite permeability the magnetic field inside the cylinder $\varrho < a$ is constant.

$$\vec{H} = \frac{1}{\mu_0}\left[\frac{\vec{e}_\varrho}{\varrho}\frac{\partial A}{\partial \varphi} - \vec{e}_\varphi\frac{\partial A}{\partial \varrho}\right] = H_e\,b\left[\vec{e}_\varrho\frac{a_1}{a}\cos\varphi - \vec{e}_\varphi\frac{a_1}{a}\sin\varphi\right]$$

$$= H_e\,b\frac{a_1}{a}\left[\vec{e}_\varrho\cos\varphi - \vec{e}_\varphi\sin\varphi\right] = \vec{e}_x\,H_e\frac{b}{a}a_1$$

$$\vec{H} = \vec{e}_x\,H_e\frac{2(1-k)}{(1+\mu_0/\mu)[1-k^2(a/b)^2]}\,.$$

Hence the screening factor $\eta_s = |\vec{H}_e|/|\vec{H}|$ is

$$\eta_s = \frac{1}{4}\frac{(\mu+\mu_0)^2}{\mu\mu_0}\left[1 - k^2\left(\frac{a}{b}\right)^2\right].$$

## 4.8   Mutual Inductance of Plane Conductor Loops

Consider two thin conductor loops of infinite extension in axial direction at positions $(\vec{r}_{11}, \vec{r}_{12})$ and $(\vec{r}_{21}, \vec{r}_{22})$. The half-space $y > 0$, that contains the conductor loops, is of permeability $\mu_0$ and the half-space $y \leq 0$ is of permeability $\mu$.

Calculate the mutual inductance of the two conductor loops per unit length.

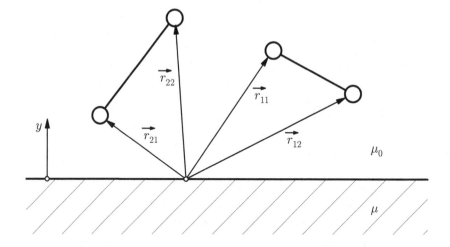

A current $I$ in the loop at positions $(\vec{r}_{11}; \vec{r}_{12})$ induces a magnetic flux $\Psi_m$ per unit length $l$ through the second loop.

$$\Psi_m/l = \int\limits_a \vec{B}\,d\vec{a}/l = \oint\limits_C \vec{A}\,d\vec{s}/l = A(\vec{r}_{21}) - A(\vec{r}_{22}); \qquad \vec{B} = \operatorname{rot}\vec{A}$$

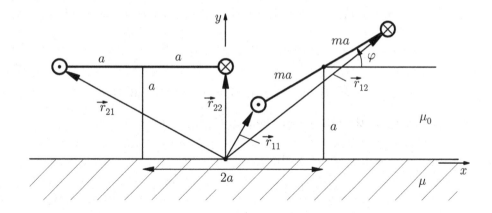

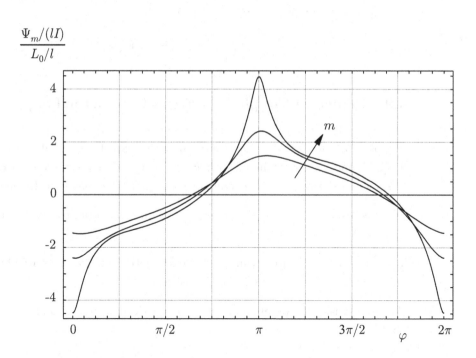

Fig. 4.8–1: Design (top) and normalized flux $\Psi_m/(lI)$ (bottom) in dependence on the angle $\varphi$

Parameters: $\mu/\mu_0 = 100$, $L_0/l = \mu/(4\pi)$, $m = 0.5;\ 0.7;\ 0.9$

The vector potential $\vec{A}(\vec{r}) = \vec{e}_z A(\vec{r})$ can be calculated from the currents $\pm I$ of the exciting loop at positions $(\vec{r}_{11}; \vec{r}_{12})$ and the image currents $kI$ at $\vec{r}_{11}^* = \vec{r}_{11} - 2\vec{e}_y(\vec{e}_y \vec{r}_{11})$ and $-kI$ at $\vec{r}_{12}^* = \vec{r}_{12} - 2\vec{e}_y(\vec{e}_y \vec{r}_{12})$ in the homogeneous space of permeability $\mu_0$.

$$A = -\frac{\mu_0 I}{2\pi}\left[\ln\frac{|\vec{r} - \vec{r}_{11}|}{|\vec{r} - \vec{r}_{12}|} + k\ln\frac{|\vec{r} - \vec{r}_{11}^*|}{|\vec{r} - \vec{r}_{12}^*|}\right] \;; \qquad k = \frac{\mu - \mu_0}{\mu + \mu_0}$$

The image currents describe the influence of the half-space $z < 0$ with permeability $\mu$.

Therefore the flux trough the loop $(\vec{r}_{21}, \vec{r}_{22})$ is

$$\begin{aligned}
\Psi_m/l &= A(\vec{r}_{21}) - A(\vec{r}_{22}) \\
&= -\frac{\mu I}{2\pi}\left[\ln\left(\frac{|\vec{r}_{21} - \vec{r}_{11}|\,|\vec{r}_{22} - \vec{r}_{12}|}{|\vec{r}_{21} - \vec{r}_{12}|\,|\vec{r}_{22} - \vec{r}_{11}|}\right) + k\ln\left(\frac{|\vec{r}_{21} - \vec{r}_{11}^*|\,|\vec{r}_{22} - \vec{r}_{12}^*|}{|\vec{r}_{21} - \vec{r}_{12}^*|\,|\vec{r}_{22} - \vec{r}_{11}^*|}\right)\right]
\end{aligned}$$

and the mutual inductance is obtained by

$$L_{12}/l = L_{21}/l = (\Psi_m/l)/I\,.$$

## 4.9   Inductive Coupling between Conductor Loops

The plane $z = 0$ contains a conductor loop consisting of two parallel wires with distance $2a$ and infinite extension in axial direction. A second loop with contour $C$ is positioned as shown in the figure. The conductor loops should be magnetically decoupled.

What is the condition for the geometry parameters? Find the parameter $d$ if $b = 2a$ and $c = 3a$.

The conductor loops are decoupled if the magnetic flux, that is excited by the first loop and runs through the second loop, vanishes.

We assume, that the parallel conductors carry the currents $\pm I$ and thus own the vector potential

$$\vec{A} = -\vec{e}_z\,\frac{\mu I}{2\pi}\,\ln(\varrho_1/\varrho_2) = -\vec{e}_z\,\frac{\mu I}{2\pi}\,\ln\sqrt{\frac{(x - a)^2 + y^2}{(x + a)^2 + y^2}}\,.$$

Then the magnetic flux

$$\Psi_m = \int_a \vec{B}d\vec{a} = \int_a \mathrm{rot}\,\vec{A}\,d\vec{a} = \oint_C \vec{A}\,d\vec{s}$$

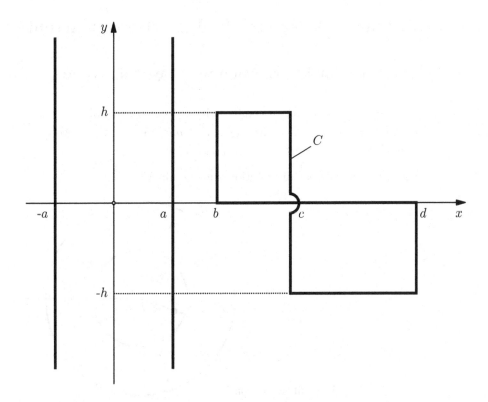

through the second loop is

$$\Psi_m = -\frac{\mu I}{2\pi}\left[h\ln\frac{b-a}{b+a} + h\ln\frac{d-a}{d+a} - 2h\ln\frac{c-a}{c+a}\right]$$

$$= -\frac{\mu I h}{2\pi}\left[\ln\frac{(b-a)(d-a)}{(b+a)(d+a)} - \ln\left(\frac{c-a}{c+a}\right)^2\right].$$

The conducting loops are decoupled if $\Psi_m = 0$, that is

$$\frac{(b-a)(d-a)}{(b+a)(d+a)} = \left(\frac{c-a}{c+a}\right)^2.$$

Finally with $b = 2a$ and $c = 3a$ we get $d = 7a$.

# 5. Quasi Stationary Fields – Eddy Currents

## 5.1   Current Distribution in a Layered Cylinder

A cylindrical conductor of conductivity $\kappa_1$ and permeability $\mu_1$ in $\varrho < a$ and of conductivity $\kappa_2$ and permeability $\mu_2$ in $a < \varrho < b$ is electrically connected in $\varrho = a$ and carries the total current $i(t) = \mathrm{Re}\{i_0 \exp(j\omega t)\}$.

Calculate the current distribution and the magnetic field.

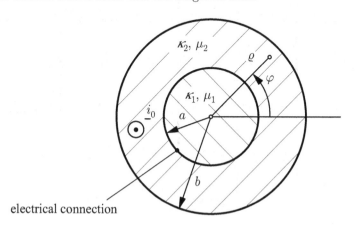

electrical connection

The field is described by the complex amplitude $\underline{\vec{A}}$ of an axial vector potential.

$$\underline{\vec{A}} = \vec{e}_z \underline{A}(\varrho) ; \quad \underline{\vec{B}} = \mathrm{rot}\,\underline{\vec{A}} ; \quad \Delta \underline{A} = \frac{d^2 \underline{A}}{d\varrho^2} + \frac{1}{\varrho}\frac{d\underline{A}}{d\varrho} = \alpha^2 \underline{A} ; \quad \alpha^2 = j\omega\kappa\mu$$

Inside the conducting material the solution of the differential equation requires modified Bessel functions.

$$\underline{A}(\varrho) = \begin{cases} \underline{C}_1 \dfrac{I_0(\alpha_1 \varrho)}{I_0(\alpha_1 a)} & ; \quad \varrho < a \\[2ex] \underline{C}_2 \dfrac{I_0(\alpha_2 \varrho)}{I_0(\alpha_2 a)} + \underline{C}_3 \dfrac{K_0(\alpha_2 \varrho)}{K_0(\alpha_2 a)} & ; \quad a < \varrho < b \end{cases}$$

$$\alpha_1^2 = j\omega\kappa_1\mu_1 ; \qquad \alpha_2^2 = j\omega\kappa_2\mu_2 = \frac{2j}{\delta^2}$$

$$\underline{\vec{H}} = \frac{1}{\mu}\,\mathrm{rot}\,[\vec{e}_z \underline{A}] = -\vec{e}_\varphi \frac{1}{\mu}\frac{\partial \underline{A}}{\partial \varrho} = \vec{e}_\varphi \underline{H}(\varrho) ; \quad \underline{\vec{E}} = -j\omega\underline{\vec{A}} ; \quad \underline{\vec{J}} = \kappa\underline{\vec{E}}$$

With the boundary conditions

$$2\pi b \underline{H}(\varrho = b) = i_0 ; \quad \underline{E}(\varrho)\big|_{\substack{\varrho<a \\ \varrho \to a}} = \underline{E}(\varrho)\big|_{\substack{\varrho>a \\ \varrho \to a}} ; \quad \underline{H}(\varrho)\big|_{\substack{\varrho>a \\ \varrho \to a}} = \underline{H}(\varrho)\big|_{\substack{\varrho<a \\ \varrho \to a}}$$

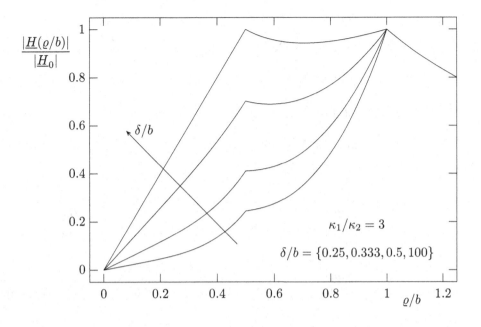

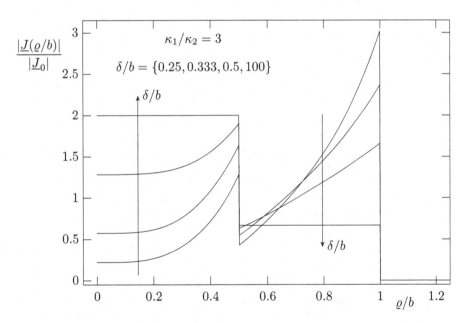

Fig. 5.1–1: Magnitude of the magnetic field (top) and current density (bottom) for different normalized skin depths $\delta/b$ and $b/a = 2$
$$\left(\underline{H}_0 = \underline{i}_0/(2\pi b), \ \underline{J}_0 = \underline{i}_0/(\pi b^2)\right)$$

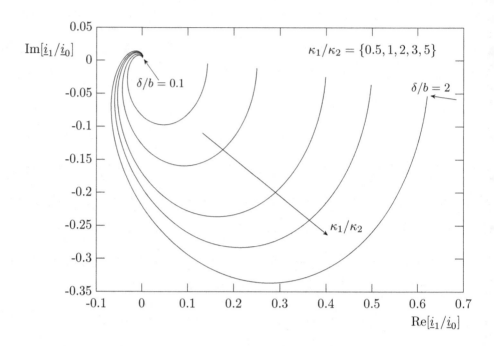

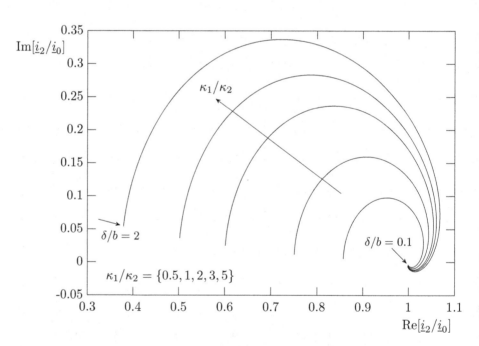

Fig. 5.1–2: Parametric plot of the normalized currents $\underline{i}_1/\underline{i}_0$ (top) and $\underline{i}_2/\underline{i}_0$ (bottom) in dependence on the normalized skin depth $\delta/b$ for different conductivity ratios $\kappa_1/\kappa_2$

the following equations for the unknown constants are derived.

$$\underline{C}_1 = \underline{C}_2 + \underline{C}_3; \quad I_0' = I_1; \quad K_0' = -K_1$$

$$\underline{C}_1 \frac{I_1(\alpha_1 a)}{I_0(\alpha_1 a)} = \frac{\mu_1}{\mu_2} \frac{\alpha_2}{\alpha_1} \left[ \underline{C}_2 \frac{I_1(\alpha_2 a)}{I_0(\alpha_2 a)} - \underline{C}_3 \frac{K_1(\alpha_2 a)}{K_0(\alpha_2 a)} \right]$$

$$-\frac{\mu_2 i_0}{2\pi \alpha_2 b} = \underline{C}_2 \frac{I_1(\alpha_2 b)}{I_0(\alpha_2 a)} - \underline{C}_3 \frac{K_1(\alpha_2 b)}{K_0(\alpha_2 a)}$$

The constants follow from a straightforward solution of the equations.

Outside the conductor the field is given by

$$\underline{H}(\varrho) = \frac{i_0}{2\pi \varrho}; \qquad \varrho \geq b.$$

Finally the currents $i_1$ and $i_2$ in the conductors $\varrho < a$ and $a < \varrho < b$ are

$$i_1 = 2\pi a \, \underline{H}(a) = -\frac{2\pi a}{\mu_1} \frac{\partial \underline{A}}{\partial \varrho} \bigg|_{\varrho = a} = -2\pi a \frac{\alpha_1}{\mu_1} \underline{C}_1 \frac{I_1(\alpha_1 a)}{I_0(\alpha_1 a)}; \qquad i_2 = i_0 - i_1.$$

## 5.2 Rotating Conductor Loop

A thin conductor loop with rectangular contour $C$ and dimensions $2a$ and $b$ rotates with angular velocity $\vec{\omega}(t) = \vec{e}_z \omega_0 \, t/T$ around its axis ($x = c < a$; $y = 0$). In $x \leq 0$ the loop is exposed to a homogeneous magnetic field $\vec{H} = \vec{e}_y H$. The whole space is of permeability $\mu$.

What is the voltage $u(t)$ measured by a highly resistive voltmeter in the time interval, when the loop rotates by an angle of $180°$ from its initial position in the plane $x = c$ at the time $t = 0$?

The induced voltage results from a contour integration of the vector $\vec{v} \times \vec{B}$, where $\vec{v} = \vec{\omega} \times \vec{r}$ is the velocity of the conductor element at position $\vec{r}$.

$$u(t) = \oint_C \vec{v} \times \vec{B} \, d\vec{s} = \mu \oint_C (\vec{\omega} \times \vec{r}) \times \vec{H} \, d\vec{s} = \mu \oint_C [\vec{r}\underbrace{(\vec{\omega}\vec{H})}_{=0} - \vec{\omega}(\vec{r}\vec{H})] \, d\vec{s}$$

$$= -\mu \int_{-b}^{0} \vec{e}_z \omega_0 t/T \, (\vec{r}(\vec{e}_y H)) \, (\vec{e}_z dz) = -\mu H \omega_0 \, t/T \int_{-b}^{0} \vec{r}\vec{e}_y \, dz = -\mu H \, ab \, \omega_0 \, t/T \, \cos(\varphi(t))$$

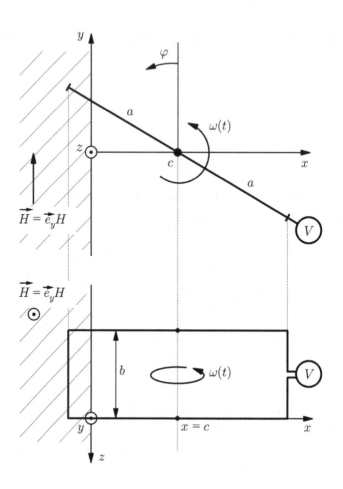

$$\varphi(t) = \int\limits_0^t \omega(t')\, dt' = \omega_0\, \frac{t^2}{2T}$$

$$u(t) = -\underbrace{\mu H a b}_{\Psi_m}\, \omega_0\, t/T\, \cos(\omega_0\, t^2/(2T)) ; \qquad t_1 < t < t_2$$

$$\varphi(t_1) = \arcsin\left(c/a\right) = \frac{\omega_0}{2}\, \frac{t_1^2}{T} ; \qquad t_1 = \left[\frac{2T}{\omega_0}\, \arcsin\left(c/a\right)\right]^{1/2}$$

$$\varphi(t_2) = \pi - \arcsin\left(c/a\right) = \frac{\omega_0}{2}\, \frac{t_2^2}{T} ; \qquad t_2 = \left[\frac{2T}{\omega_0}\, \left(\pi - \arcsin\left(c/a\right)\right)\right]^{1/2}$$

The result for $t < \sqrt{2\pi\,T/\omega_0}$ is

$$
u(t) = \begin{cases}
0 & ; \quad 0 < t < t_1 \\
-\Psi_m\,\omega_0 t/T\,\cos(\omega_0 t^2/(2T)) & ; \quad t_1 < t < t_2 \\
0 & ; \quad t_2 < t < \sqrt{2\pi T/\omega_0}\,.
\end{cases}
$$

## 5.3  Force Caused by an Induced Current Distribution inside a Conducting Sphere

A thin conductor loop that carries the current $i = i_0\cos(\omega t + \varphi) = \mathrm{Re}\{i_0\exp(j\omega t)\}$ is positioned at $(r = b;\ \vartheta = \vartheta_0)$ in front of a sphere with radius $a < b$ and conductivity $\kappa$. The permeability $\mu$ is constant.

Calculate the induced current distribution and determine the force acting on the conductor loop. Use the result to analyze the limit cases, where the exciting field is homogeneous and where the source is a magnetic dipole. Furthermore calculate the limit of a high-frequency stimulation $(\omega \to \infty)$.

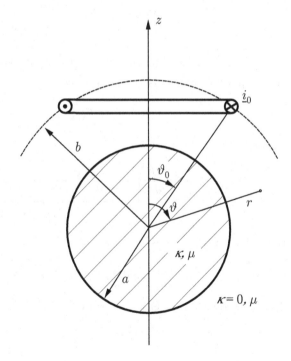

**Differential Equation and Solution Functions**

The complex amplitude of the vector potential is $\varphi$-directed throughout the space

$$\vec{\underline{B}} = \text{rot}\,\vec{\underline{A}}\,; \qquad \vec{\underline{A}} = \vec{e}_\varphi\,\underline{A}(r,\vartheta)\,; \qquad \frac{\partial \underline{A}}{\partial \varphi} = 0.$$

In $r < a$ it must satisfy the skin effect equation

$$\text{rot}\,\text{rot}\,(\vec{e}_\varphi\underline{A}) = -\kappa\mu\frac{\partial}{\partial t}\,\vec{e}_\varphi\underline{A}\,; \qquad \text{rot}\,\text{rot}\,(\vec{e}_\varphi\underline{A}) = -j\omega\kappa\mu\vec{e}_\varphi\underline{A}$$

and in $r > a, r \neq b$ because of $\kappa = 0$ the equation

$$\text{rot}\,\text{rot}\,(\vec{e}_\varphi\underline{A}) = 0.$$

With spherical coordinates $(r,\vartheta,\varphi)$ the skin effect equation for $\underline{A}$ reads

$$r^2\frac{\partial^2\underline{A}}{\partial r^2} + 2r\frac{\partial\underline{A}}{\partial r} + \frac{\partial}{\partial\vartheta}\left[\frac{1}{\sin\vartheta}\frac{\partial}{\partial\vartheta}(\sin\vartheta\,\underline{A})\right] = j\omega\kappa\mu\underline{A}\cdot r^2 = (\alpha r)^2\underline{A}\,,$$

where the curl operator $\text{rot}\,\vec{F}$ in spherical coordinates

$$\text{rot}\,\vec{F} = \frac{1}{r^2\sin\vartheta}\begin{vmatrix} \vec{e}_r & \vec{e}_\vartheta\,r & \vec{e}_\varphi\,r\,\sin\vartheta \\ \partial/\partial r & \partial/\partial\vartheta & \partial/\partial\varphi \\ F_r & r\,F_\vartheta & r\,\sin\vartheta\,F_\varphi \end{vmatrix}$$

has been applied two times on the vector $\vec{\underline{A}} = \vec{e}_\varphi\,\underline{A}$. We try to find a solution of the differential equation by means of the product:

$$\underline{A} = \underline{R}(r)\cdot\Theta(\vartheta)\,.$$

This leads to

$$\frac{1}{\underline{R}}\left(r^2\frac{d^2\underline{R}}{dr^2} + 2r\frac{d\underline{R}}{dr}\right) + \underbrace{\frac{1}{\Theta}\frac{d}{d\vartheta}\left[\frac{1}{\sin\vartheta}\frac{d}{d\vartheta}(\Theta\sin\vartheta)\right]}_{-n(n+1)} - (\alpha r)^2 = 0\,.$$

With the substitution $u = \cos\vartheta$

$$\frac{d}{d\vartheta} = \frac{d}{du}\frac{du}{d\vartheta} = -\sin\vartheta\frac{d}{du} = -\sqrt{1-u^2}\frac{d}{du}\,; \qquad n = 0,1,2,\ldots$$

the differential equation for $\Theta(u)$ becomes

$$\frac{d}{du}\left[(1-u^2)\frac{d\Theta}{du}\right] + \left[n(n+1) - \frac{1}{1-u^2}\right]\Theta = 0\,.$$

This is a special case of the more general differential equation

$$\frac{d}{du}\left[(1-u^2)\frac{d\Theta}{du}\right] + \left[n(n+1) - \frac{m^2}{1-u^2}\right]\Theta = 0.$$

The solutions are the spherical harmonics $P_n^m(u)$ and $Q_n^m(u)$, that restrict to $m = 1$ in this case. Furthermore the functions of second kind $Q_n^1(u)$ must be omitted as they increase unbounded for points $u = \pm 1$, i.e. $\vartheta = 0, \pi$, which are part of the computational domain.

For the function $\underline{R}(r)$ the following differential equation remains.

$$r^2\frac{d^2\underline{R}}{dr^2} + 2r\frac{d\underline{R}}{dr} - \left[n(n+1) + (\alpha r)^2\right]\underline{R} = 0; \qquad \alpha^2 = j\omega\kappa\mu = 2j/\delta^2$$

Its solution are the modified spherical Bessel functions

$$R_n(\alpha r) = \sqrt{\frac{\pi}{2\alpha r}}I_{n+\frac{1}{2}}(\alpha r) \qquad \text{and} \qquad T_n(\alpha r) = \sqrt{\frac{\pi}{2\alpha r}}K_{n+\frac{1}{2}}(\alpha r).$$

Here only the functions $R_n(\alpha r)$ have to be regarded, because the functions $T_n(\alpha r)$ are unbounded for $r \to 0$.

Thus, all solution functions of the skin effect equation for a $\varphi$-directed vector potential inside a conducting sphere are known.

In $r > a$ outside the sphere $\alpha^2 = 0$ holds. The differential equation for the function $\Theta(\vartheta)$ applies unmodified and the differential equation for $\underline{R}$ becomes

$$r^2\frac{d^2\underline{R}}{dr^2} + 2r\frac{d\underline{R}}{dr} - n(n+1)\underline{R} = 0.$$

Solutions of this equation are the functions $r^n$ and $r^{-(n+1)}$.

**Exciting Potential in the Homogeneous Space**

A first step to the solution of the current problem is the determination of the exciting vector potential $\vec{A}_e = \vec{e}_\varphi \underline{A}_e(r, \vartheta)$ of the conductor loop at position $(r = b; \vartheta = \vartheta_0)$, when the surrounding space is homogeneous and of permeability $\mu$.

For this purpose we replace the loop by a current sheet $\underline{K}(\vartheta)$ on the sphere $r = b$, that is $\varphi$-directed and depends on the coordinate $\vartheta$. In the surrounding medium outside of the surface $r = b$ the vector potential must satisfy

$$\text{rot}\,\text{rot}\,(\vec{e}_\varphi\,\underline{A}_e) = 0,$$

thus with the considerations of the previous section the solution is:

$$\underline{A}_e = \sum_{n=1}^{\infty}\left\{\begin{array}{ll}\underline{B}_n & (b/r)^{n+1} \\ \underline{C}_n & (r/b)^n\end{array}\right\}P_n^1(\cos\vartheta)\,; \qquad \begin{array}{l} r > b \\ r < b \end{array}.$$

Here the solution $n = 0$ vanishes, because of $P_0^1 = 0$.

In the boundary layer $r = b$ the normal component of the magnetic flux density is continuous and as a consequence, the potential itself is continuous. Thus

$$\underline{B}_n = \underline{C}_n .$$

A second condition follows from the boundary condition for the tangential component of the magnetic field $\vec{\underline{H}}_e = 1/\mu \vec{\underline{B}}_e = 1/\mu \operatorname{rot} \vec{\underline{A}}_e$.

$$\vec{e}_\vartheta \operatorname{rot} \left( \vec{\underline{A}}_e \Big|_{r>b} - \vec{\underline{A}}_e \Big|_{r<b} \right) \Bigg|_{r \to b} = \mu \underline{K}(\vartheta)$$

With the previously given expression for the curl-operator in spherical coordinates it follows

$$\frac{1}{r} \frac{\partial}{\partial r} \left[ r \left( \underline{A}_e|_{r<b} - \underline{A}_e|_{r>b} \right) \right] \Big|_{r \to b} = \mu \underline{K}(\vartheta)$$

and with the expressions for $\underline{A}_e$

$$\sum_{n=1}^{\infty} \underline{B}_n \left[ \frac{n}{r} \left( \frac{b}{r} \right)^{n+1} + \frac{n+1}{r} \left( \frac{r}{b} \right)^n \right]_{r=b} P_n^1(\cos \vartheta) = \mu \underline{K}(\vartheta) .$$

Now multiplication with $P_i^1(\cos \vartheta) \sin \vartheta$ and integration leads to

$$\sum_{n=1}^{\infty} (2n+1) \underline{B}_n \int_0^\pi P_n^1(\cos \vartheta) P_i^1(\cos \vartheta) \sin \vartheta \, d\vartheta = \int_0^\pi \mu b \underline{K}(\vartheta) P_i^1(\cos \vartheta) \sin \vartheta \, d\vartheta$$

and due to the orthogonality of the spherical harmonics

$$\int_0^\pi P_n^m(\cos \vartheta) P_i^m(\cos \vartheta) \sin \vartheta \, d\vartheta = \begin{cases} \dfrac{(n+m)!}{(n-m)!} \dfrac{2}{2n+1} & ; \quad i = n \\ 0 & ; \quad i \neq n \end{cases}$$

the result is

$$(2n+1) \underline{B}_n \frac{(n+1)!}{(n-1)!} \frac{2}{2n+1} = \mu P_n^1(\cos \vartheta_0) \sin \vartheta_0 \underbrace{\int_0^\pi \underline{K}(\vartheta) b \, d\vartheta}_{\underline{i}_0} .$$

The current sheet $\underline{K}(\vartheta)$ vanishes except for $\vartheta = \vartheta_0$ and thus the integral on the right yields the current $\underline{i}_0$. Hence the constants are determined.

$$\underline{C}_n = \underline{B}_n = \frac{\mu \underline{i}_0 \sin \vartheta_0 P_n^1(\cos \vartheta_0)}{2n(n+1)}$$

### Field in Presence of the Conducting Sphere

After inserting the conducting sphere the vector potential of the induced current distribution must be superposed to the exciting potential $\underline{A}_e$. In the preliminary section the following solution has been derived for $r < a$:

$$\underline{A} = \frac{\mu \, \underline{i}_0 \, \sin \vartheta_0}{2} \sum_{n=1}^{\infty} \underline{D}_n \, R_n(\alpha r) \, P_n^1(\cos \vartheta_0) \, P_n^1(\cos \vartheta) \, .$$

Here $\underline{D}_n$ is a suitably chosen constant. In $r > a$ applies

$$\underline{A} = \underline{A}_e + \frac{\mu \underline{i}_0 \, \sin \vartheta_0}{2} \sum_{n=1}^{\infty} \underline{E}_n \left( \frac{b}{r} \right)^{n+1} P_n^1(\cos \vartheta_0) P_n^1(\cos \vartheta) = \underline{A}_e + \underline{A}_S \, .$$

This approach already satisfies all boundary conditions in $r = b$. The evaluation of the boundary conditions in $r = a$ leads to the new constants $\underline{D}_n$ and $\underline{E}_n$.

From the continuity of the normal component of the magnetic flux density in $r = a$ follows the continuity of the vector potential.

$$\underline{D}_n \, R_n \, (\alpha a) = \frac{(a/b)^n}{n(n+1)} + \underline{E}_n \, (b/a)^{n+1}$$

A second equation follows from the continuity of the tangential component of the magnetic field

$$\left( \frac{\partial}{\partial r} (r \, \underline{A}) \bigg|_{r<a} - \frac{\partial}{\partial r} (r \, \underline{A}) \bigg|_{r>a} \right) \bigg|_{r \to a} = 0$$

with the result

$$\underline{D}_n \, [R_n \, (\alpha a) + \alpha a \, R_n' \, (\alpha a)] = \underline{D}_n \, [(n+1) \, R_n + \alpha a \, R_{n+1}] = \frac{1}{n} \left( \frac{a}{b} \right)^n - n \left( \frac{b}{a} \right)^{n+1} \underline{E}_n \, .$$

Here it is

$$\alpha r \, R_n'(\alpha r) = [n \, R_n \, (\alpha r) + \alpha r \, R_{n+1} \, (\alpha r)]$$

and the recurrence relation reads

$$(2n+1) \, R_n \, (\alpha r) + \alpha r \, R_{n+1} \, (\alpha r) = \alpha r \, R_{n-1}(\alpha r) \, .$$

Now, multiplying the first equation by $n$ and adding the second equation yields

$$\underline{D}_n \underbrace{[(2n+1) \, R_n \, (\alpha a) + \alpha a \, R_{n+1} \, (\alpha a)]}_{\alpha a \, R_{n-1} \, (\alpha a)} = \frac{2n+1}{n(n+1)} \left( \frac{a}{b} \right)^n$$

$$\underline{D}_n = \frac{2n+1}{n(n+1)} \frac{(a/b)^n}{\alpha a \, R_{n-1} \, (\alpha a)} \, ; \qquad \underline{E}_n = - \frac{(a/b)^{2n+1}}{n(n+1)} \frac{R_{n+1} \, (\alpha a)}{R_{n-1} \, (\alpha a)} \, .$$

Hence the vector potential in $r \leq a$ is

$$\underline{A} = \frac{\mu \, \underline{i}_0 \, \sin \vartheta_0}{2} \sum_{n=1}^{\infty} \frac{2n+1}{n(n+1)} \frac{R_n(\alpha r)}{\alpha a \, R_{n-1}(\alpha a)} \left(\frac{a}{b}\right)^n P_n^1(\cos \vartheta_0) \, P_n^1(\cos \vartheta) ; \quad r \leq a .$$

In this problem the current density $\vec{J}$ can be derived by the time derivative of the vector potential.

$$\underline{\vec{J}} = -j\omega\kappa\underline{\vec{A}} = \vec{e}_\varphi \, \underline{J}$$

$$\underline{J}(r, \vartheta) = -\frac{1}{2} \frac{\underline{i}_0}{a^2} \alpha a \sin \vartheta_0 \sum_{n=1}^{\infty} \frac{2n+1}{n(n+1)} \frac{R_n(\alpha r)}{R_{n-1}(\alpha a)} \left(\frac{a}{b}\right)^n P_n^1(\cos \vartheta_0) P_n^1(\cos \vartheta)$$

In the limit $\omega \to 0$ it follows with the asymptotic expansion for spherical Bessel functions

$$R_n(\alpha r) \big|_{\alpha \to 0} \approx \frac{(\alpha r)^n}{1 \cdot 3 \cdot 5 \cdots (2n+1)}$$

$$\frac{R_n(\alpha r)}{\alpha a \, R_{n-1}(\alpha a)} \bigg|_{\alpha \to 0} = \frac{(\alpha r)^n}{(2n+1)(\alpha a)^n} \bigg|_{\alpha \to 0} = \frac{1}{2n+1} \left(\frac{r}{a}\right)^n .$$

Thus the vector potential is

$$\underline{A} \big|_{\omega \to 0} = \frac{\mu \underline{i}_0 \sin \vartheta_0}{2} \sum_{n=1}^{\infty} \frac{1}{n(n+1)} \left(\frac{r}{b}\right)^n P_n^1(\cos \vartheta_0) \, P_n^1(\cos \vartheta) = \underline{A}_e .$$

As expected only the exciting potential remains. The differential equation for the magnetic lines of force at the time $t = 0$ is

$$d\vec{s} \times \mathrm{Re}\{\underline{\vec{B}}\} = d\vec{s} \times \mathrm{rot} \, [\vec{e}_\varphi \mathrm{Re}\{\underline{A}(r, \vartheta)\}] = 0 .$$

However, in the present case it is easier to evaluate the magnetic flux

$$\underline{\Psi}_m = \int_a \underline{\vec{B}} d\vec{a} = \int_a \mathrm{rot} \, \underline{\vec{A}} \, d\vec{a} = \oint_C \underline{\vec{A}} \, d\vec{s} = \mathrm{const} .$$

Because of the rotational symmetry the closed paths $C$ are concentric circles around the $z$-axis and we obtain the equation

$$\Psi_m = r \sin \vartheta \, \mathrm{Re} \, \{\underline{A}(r, \vartheta)\} = \mathrm{const}$$

for the magnetic lines of force.

## Sphere Excited by a Homogeneous Field $\underline{H}_0$

If the conductor loop is positioned in the plane $\vartheta_0 = \pi/2$ and if its radius is large compared to the radius of the sphere $a/b \ll 1$, with $i_0/b = \text{const}$, then the expression of the vector potential reduces to the first summation term:

$$\underline{J}(r,\vartheta) = -\frac{3}{4}\frac{i_0}{ab}\sin\vartheta_0\,\alpha a\,\frac{R_1(\alpha r)}{R_0(\alpha a)}\,P_1^1\left(\cos\vartheta_0\right)P_1^1\left(\cos\vartheta\right)$$

$$\text{with}\quad \vartheta_0 = \frac{\pi}{2}\,;\qquad P_1^1\left(\cos\vartheta\right)=\sin\vartheta\qquad\text{and}\qquad \underline{H}_0 = \frac{i_0}{2b}$$

$$\underline{J}(r,\vartheta) = -\frac{3}{2}\frac{\underline{H}_0}{a}\,\alpha a\frac{R_1(\alpha r)}{R_0(\alpha a)}\sin\vartheta\,.$$

## Excitation by a Magnetic Dipole

For small angles $\vartheta_0$ the conductor loop can be replaced by a magnetic dipole of moment

$$\vec{M} = \vec{e}_z\,\mu\,\underline{i}_0\,\pi(b\sin\vartheta_0)^2 = \vec{e}_z\underline{M}\,.$$

With

$$P_n^1\left(\cos\vartheta_0\right) = \sin\vartheta_0\,P_n'\left(\cos\vartheta_0\right)\,;\qquad P_n'(1) = \frac{1}{2}n(n+1)$$

it follows

$$\underline{J}(r,\vartheta) = -\frac{\underline{M}}{4\pi\mu}\frac{\alpha a}{(ab)^2}\sum_{n=1}^{\infty}(2n+1)\left(\frac{a}{b}\right)^n\frac{R_n(\alpha r)}{R_{n-1}(\alpha a)}P_n^1(\cos\vartheta)\,.$$

## The High Frequency Limit

In the limit $\omega \to \infty$ and thus $\alpha^2 = j\omega\kappa\mu \to \infty$ the constants $\underline{E}_n$ in the vector potential for points $r \geq a$ are given by

$$\underline{E}_n\big|_{\omega\to\infty} = -\frac{1}{n(n+1)}\left(\frac{a}{b}\right)^{2n+1}\,.$$

The vector potential in $r \geq a$ becomes

$$\underline{A}\big|_{\omega\to\infty} = \frac{\mu\,\underline{i}_0\,\sin\vartheta_0}{2}\sum_{n=1}^{\infty}\frac{1}{n(n+1)}\left[\left(\frac{r}{b}\right)^n - \left(\frac{a}{b}\right)^n\left(\frac{a}{r}\right)^{n+1}\right]P_n^1\left(\cos\vartheta_0\right)P_n^1(\cos\vartheta)\,.$$

Now the current sheet on the sphere follows from

$$\vec{n}\times\vec{\underline{H}}\Big|_{r=a} = \vec{\underline{K}} = \vec{e}_\varphi\,\underline{K}\left(\vartheta\right)$$

with the result

$$\underline{K}(\vartheta) = -\frac{1}{\mu}\frac{1}{r}\frac{\partial}{\partial r}\left(r\underline{A}\right)\big|_{r>a,r\to a,\omega\to\infty}$$

$$= -\frac{\underline{i}_0\,\sin\vartheta_0}{2a}\sum_{n=1}^{\infty}\frac{2n+1}{n(n+1)}\left(\frac{a}{b}\right)^n P_n^1(\cos\vartheta_0)\,P_n^1(\cos\vartheta)\,.$$

## Force Acting on the Conductor Loop

In general the force on a conductor of volume $v$ and current density $\vec{J}$ is calculated by

$$\vec{F} = \int_v \vec{J} \times \vec{B} \, dv.$$

Due to the symmetry we can write in the present case

$$\vec{F} = i(t) \oint_C d\vec{s} \times \vec{B}_S(\vec{r}, t) = \vec{e}_z F(t)$$

$$F = i(t) \oint_C \vec{e}_z (d\vec{s} \times \vec{B}_S) = i(t) \oint_C (\vec{e}_z \times d\vec{s}) \vec{B}_S = -i(t) 2\pi b \sin \vartheta_0 \, B_{S\varrho}|_{\vartheta=\vartheta_0; r=b}.$$

The flux density $B_{S\varrho}(t)$ is solely excited by the induced eddy currents, which follow from the vector potential $\vec{\underline{A}}_S$ with

$$B_{S\varrho} = \mathrm{Re}\{\vec{e}_\varrho \mathrm{rot}\,(\vec{\underline{A}}_S) \exp(j\omega t)\}.$$

$$F(t) = -2\pi b \sin \vartheta_0 \, i(t) \left[ B_{Sr}(t) \sin \vartheta + B_{S\vartheta}(t) \cos \vartheta \right]|_{r=b; \vartheta=\vartheta_0}$$

$$= -2\pi b \sin \vartheta_0 \, i(t) \left[ \frac{1}{r} \frac{\partial}{\partial \vartheta}(\sin \vartheta A_S) - \frac{\cos \vartheta}{r} \frac{\partial}{\partial r}(r A_S) \right]_{r=b; \vartheta=\vartheta_0}$$

$$= -2\pi \sin \vartheta_0 \, i(t) \left[ \sin \vartheta \frac{\partial A_S}{\partial \vartheta} - \cos \vartheta b \frac{\partial A_S}{\partial r} \right]_{r=b; \vartheta=\vartheta_0}$$

$$A_S(t) = A(t) - A_e(t) = \frac{\mu \sin \vartheta_0}{2} \sum_{n=1}^{\infty} \mathrm{Re}\{\underline{i}_0 \underline{E}_n \exp(j\omega t)\} (b/r)^{n+1} P_n^1(\cos \vartheta) P_n^1(\cos \vartheta_0)$$

$$u = \cos \vartheta; \quad u_0 = \cos \vartheta_0$$

$$F = \frac{1}{2} \mu \pi (1 - u_0^2) \sum_{n=1}^{\infty} \mathrm{Re}\left\{ |\underline{i}_0|^2 \underline{E}_n^* + \underline{i}_0^2 \underline{E}_n \exp(2j\omega t) \right\} \cdot$$

$$\cdot \underbrace{\left[ (1 - u_0^2) \frac{dP_n^1}{du_0} - (n+1)u_0 P_n^1(u_0) \right]}_{-n P_{n+1}^1(u_0)} P_n^1(u_0) = \overline{F} + F_\sim$$

With the recurrence relation for the spherical harmonics the result reads

$$\overline{F} = -\frac{1}{2}\mu |\underline{i}_0|^2 \pi(1 - u_0^2) \sum_{n=1}^{\infty} n\,\mathrm{Re}\{\underline{E}_n\} P_{n+1}^1(u_0) P_n^1(u_0)$$

$$F_\sim = -\frac{1}{2}\mu |\underline{i}_0|^2 \pi(1 - u_0^2) \sum_{n=1}^{\infty} n\,\mathrm{Re}\left\{ \frac{\underline{i}_0^2}{|\underline{i}_0|^2} \underline{E}_n \exp(2j\omega t) \right\} P_{n+1}^1(u_0) P_n^1(u_0).$$

The force is zero if the conductor loop is located in the plane $\vartheta_0 = \pi/2$, because $P_{n+1}^1(0) P_n^1(0) = 0$.

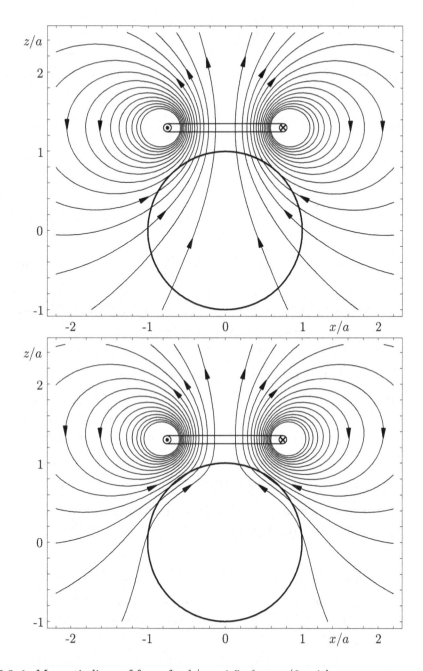

Fig. 5.3–1: Magnetic lines of force for $b/a = 1.5$, $\vartheta_0 = \pi/6$, with
$\delta/a = 1$ (top) and $\delta/a = 0.2$ (bottom)
Skin depth $\delta = \sqrt{2/(\omega\kappa\mu)}$

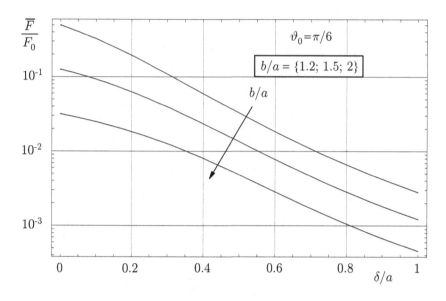

Fig. 5.3–2: Average value of the force on the conductor loop as a function of the normalized skin depth $\delta/a$ $(F_0 = \mu_0 |\underline{i}_0|^2)$

## 5.4   Impedance of a Coaxial Cable

The inner conductor of radius $a$ and the outer conductor of radii $b > a$ and $c > b$ of a coaxial cable have the conductivity $\kappa$. The permeability $\mu$ is constant.

Calculate the frequency-dependent impedance per unit length $\underline{Z}/l = R/l + j\omega L/l$ and analyze the limits of high and low frequencies.

The complex amplitude of the axially directed vector potential $\underline{\vec{A}}(\varrho) = \vec{e}_z \underline{A}(\varrho)$ satisfies the differential equation

$$\Delta \underline{A} = \alpha^2 \underline{A} \; ; \qquad \alpha^2 = j\omega\kappa\mu = \frac{2j}{\delta^2} \; .$$

The solution functions are the modified Bessel functions $I_0(\alpha\varrho)$ and $K_0(\alpha\varrho)$. As the magnetic field follows from the derivative in radial direction

$$\underline{\vec{H}} = \frac{1}{\mu} \operatorname{rot} [\vec{e}_z \underline{A}(\varrho)] = -\vec{e}_\varphi \frac{1}{\mu} \frac{\partial \underline{A}}{\partial \varrho} = \vec{e}_\varphi \underline{H}(\varrho)$$

it is described by the functions $I_1(\alpha\varrho)$ and $K_1(\alpha\varrho)$.

If the coaxial cable carries the current $\underline{i}_0$ (positive $z$-directed in the inner conductor)

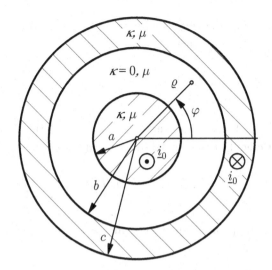

then the magnetic field in the non-conducting area $a \leq \varrho \leq b$ is

$$\underline{H}(\varrho) = \underline{H}_0 \frac{a}{\varrho} ; \qquad \underline{H}_0 = \frac{i_0}{2\pi a} .$$

Inside the inner conductor $\varrho \leq a$ the magnetic field is solely described by the function $I_1(\alpha\varrho)$ and takes the value $\underline{H}_0$ on the boundary $\varrho = a$.

$$\underline{H}(\varrho) = \underline{H}_0 \frac{I_1(\alpha\varrho)}{I_1(\alpha a)} ; \qquad \varrho \leq a$$

The field in the outer conductor is given by a linear combination of both $I_1(\alpha\varrho)$ and $K_1(\alpha\varrho)$

$$\underline{H}(\varrho) = \underline{C} \left[ I_1(\alpha\varrho) K_1(\alpha c) - K_1(\alpha\varrho) I_1(\alpha c) \right],$$

so that the field vanishes for $\varrho \geq c$. Finally the constant $\underline{C}$ follows from the condition $\underline{H}(b) = \underline{H}_0 a/b$, thus

$$\underline{H}(\varrho) = \underline{H}_0 \frac{a}{b} \frac{I_1(\alpha\varrho) K_1(\alpha c) - K_1(\alpha\varrho) I_1(\alpha c)}{I_1(\alpha b) K_1(\alpha c) - K_1(\alpha b) I_1(\alpha c)} ; \qquad b \leq \varrho \leq c.$$

Inside the conducting material the electric field is given by $\vec{\underline{J}} = \kappa \vec{\underline{E}} = \vec{e}_z \kappa \underline{E}(\varrho)$.

$$\kappa \underline{E} = \mathrm{rot}\, \vec{\underline{H}} = \vec{e}_z \frac{1}{\varrho} \frac{\partial}{\partial \varrho} \left( \varrho\, \underline{H}(\varrho) \right) ; \qquad \underline{E}(\varrho) = \frac{\alpha}{\kappa} \left[ \frac{\underline{H}(\alpha\varrho)}{\alpha\varrho} + \underline{H}'(\alpha\varrho) \right]$$

$$\underline{E}(\varrho) = \frac{\alpha}{\kappa} \underline{H}_0 \begin{cases} \dfrac{I_0(\alpha\varrho)}{I_1(\alpha a)} & ; \quad \varrho < a \\[2ex] \dfrac{a}{b} \dfrac{I_0(\alpha\varrho) K_1(\alpha c) + K_0(\alpha\varrho) I_1(\alpha c)}{I_1(\alpha b) K_1(\alpha c) - K_1(\alpha b) I_1(\alpha c)} & ; \quad b < \varrho < c \end{cases}$$

Now applying the complex Poynting theorem with $\underline{\vec{S}} = 1/2\,\underline{\vec{E}} \times \underline{\vec{H}}^*$ and integration over the conductor surfaces $a_L$ leads to

$$\overline{P}_{ve} + 2j\omega\,\overline{W}_m = -\oint_{a_L} \underline{\vec{S}}\,d\vec{a}_L = \frac{1}{2}|\underline{i}_0|^2\,R + \frac{1}{2}j\omega\mu\int_{v_L} |\underline{H}|^2\,dv_L.$$

Here $\overline{P}_{ve}$ is the time-averaged power loss and $\overline{W}_m$ is the time-averaged energy of the magnetic field in the conductor of volume $v_L$. Hence it is

$$R + j\omega L_i = -\frac{2}{|\underline{i}_0|^2}\oint_{a_L}\underline{\vec{S}}\,d\vec{a}_L$$

with the frequency-dependent inner self-inductance $L_i$ and the frequency-dependent resistance $R$.

For the present problem the result is

$$-(\overline{P}_{ve} + 2j\omega\,\overline{W}_m)/l = \pi a\big[\vec{e}_z\,\underline{E}(a) \times \vec{e}_\varphi\,\underline{H}^*(a)\big]\vec{e}_\varrho + \pi b\big[\vec{e}_z\,\underline{E}(b) \times \vec{e}_\varphi\,\underline{H}^*(b)\big](-\vec{e}_\varrho)$$

$$R/l + j\omega L_i/l = \frac{2\pi a}{|\underline{i}_0|^2}\left[\underline{E}(a)\,\underline{H}^*(a) - \frac{b}{a}\,\underline{E}(b)\,\underline{H}^*(b)\right]$$

$$= \frac{1}{2\pi a}\frac{\alpha}{\kappa}\left[\frac{I_0(\alpha a)}{I_1(\alpha a)} - \frac{a}{b}\frac{I_0(\alpha b)\,K_1(\alpha c) + K_0(\alpha b)\,I_1(\alpha c)}{I_1(\alpha b)\,K_1(\alpha c) - K_1(\alpha b)\,I_1(\alpha c)}\right].$$

For small frequencies ($|z| \ll 1$) the approximations

$$I_0(z) \approx 1;\qquad I_1(z) \approx z/2;\qquad K_0(z) \approx -\ln z;\qquad K_1(z) \approx \frac{1}{z}$$

hold an the resistance per unit length is

$$R/l\big|_{\omega \to 0} = \frac{1}{\kappa\pi a^2}\left[1 + \frac{a^2}{c^2 - b^2}\right] = R_0/l.$$

For high frequencies ($|z| \gg 1$) one can make use of the asymptotic expansions of the modified Bessel functions

$$I_n(z) \approx \sqrt{\frac{1}{2\pi z}}\,\exp(z);\qquad K_n(z) \approx \sqrt{\frac{\pi}{2z}}\,\exp(-z)$$

with the result

$$R/l + j\omega L_i/l\big|_{\omega \to \infty} \approx \frac{1+j}{2\pi\kappa a\delta}\left[1 + \frac{a}{b}\right];\qquad \alpha a = (1+j)\frac{a}{\delta}.$$

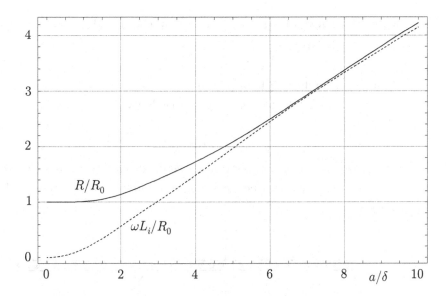

Fig. 5.4–1: Real- and imaginary part of the impedance as a function of the normalized
reciprocal skin depth $a/\delta$ for $b/a = 1.5$

The frequency-dependent outer self-inductance $L_a/l$ per unit length follows from the
time-averaged energy $\overline{W}_{ma}/l$ of the field in the non-conducting area. The integral of
the energy density over the cross-section $a_q$ gives

$$1/4 L_a/l |\underline{i}_0|^2 = \overline{W}_{ma}/l = \frac{1}{4} \int\limits_{a_q} \mathrm{Re}\left\{ \vec{\underline{H}}\,\vec{\underline{B}}^* \right\} da_q =$$

$$= \frac{1}{2}\pi\mu \int\limits_{a}^{b} |\underline{H}(\varrho)|^2 \varrho\, d\varrho = \frac{1}{2}\pi\mu\, |\underline{H}_0|^2\, a^2 \int\limits_{a}^{b} \frac{d\varrho}{\varrho}$$

$$\Rightarrow\quad L_a/l = \frac{\mu}{2\pi}\ln(b/a)\,.$$

Finally the impedance per unit length is given by

$$\underline{Z}/l = R/l + j\omega L/l\,;\qquad L = L_a + L_i\,.$$

## 5.5   Induced Current Distribution in the Conducting Half-Space

Consider a conducting half-space $y \leq 0$ of permeability $\mu$ and conductivity $\kappa$. In the non-conducting half-space $y > 0$ of permeability $\mu_0$ the plane $y = c$ carries the current sheet

$$\vec{K} = \vec{e}_z \, K_0 \cos(\pi x/a) \cos(\omega t + \varphi).$$

Find the induced current distribution and calculate the time-averaged power loss in a section $2a$ on the $x$-axis per unit length in $z$-direction. Analyze the limits $\omega \to 0$ and $\omega \to \infty$.

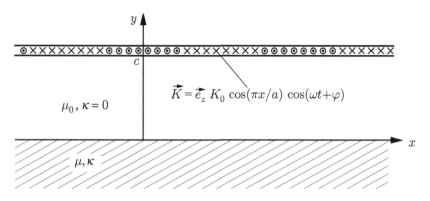

The field is described by a $z$-directed vector potential $\vec{A} = \vec{e}_z \, A(x, y)$, that satisfies the differential equation

$$\Delta \underline{A} = \frac{\partial^2 \underline{A}}{\partial x^2} + \frac{\partial^2 \underline{A}}{\partial y^2} = \begin{cases} \alpha^2 \underline{A} & ; \quad y < 0 \\ 0 & ; \quad y > 0; \quad y \neq c \end{cases} \quad ; \quad \alpha^2 = j\omega\kappa\mu = \frac{2j}{\delta^2}.$$

At first the complex potential $\vec{\underline{A}}_e(x, y) = \vec{e}_z \, \underline{A}_e(x, y)$ is calculated, that is excited, when the surrounding space of the current sheet

$$\underline{K}(x) = K_0 \exp(j\varphi) \cos(\pi x/a) = \underline{K}_0 \cos(\pi x/a)$$

is homogeneous with permeability $\mu_0$. The solution of the differential equation

$$\Delta \underline{A}_e(x, y) = \frac{\partial^2 \underline{A}_e}{\partial x^2} + \frac{\partial^2 \underline{A}_e}{\partial y^2} = 0 ; \qquad y \neq c$$

is according to the excitation given by

$$\underline{A}_e(x, y) = \underline{C} \, \cos(\pi x/a) \exp(-\pi|y - c|/a).$$

From this, the magnetic field $\underline{\vec{H}}_e = \vec{e}_x \underline{H}_{ex} + \vec{e}_y \underline{H}_{ey}$ is derived by

$$\underline{\vec{H}}_e = \frac{1}{\mu_0} \operatorname{rot} [\vec{e}_z \underline{A}_e(x, y)] = \frac{1}{\mu_0} \left[ \vec{e}_x \frac{\partial \underline{A}_e}{\partial y} - \vec{e}_y \frac{\partial \underline{A}_e}{\partial x} \right] .$$

The evaluation of the boundary condition

$$[\underline{H}_{ex}(x, y < c) - \underline{H}_{ex}(x, y > c)]_{y \to c} = \underline{K}_0 \cos(\pi x/a)$$

leads to the constant

$$\underline{C} = \frac{\mu_0 a \underline{K}_0}{2\pi} .$$

In the presence of the conducting half-space the resulting vector potential $\underline{A}(x, y)$ must satisfy the differential equation $\Delta \underline{A} = \alpha^2 \underline{A}$ in $y \leq 0$. An appropriate approach is

$$\underline{A}(x, y) = \frac{\mu_0 a \underline{K}_0}{2\pi} \cos\left(\pi \frac{x}{a}\right) \begin{cases} \exp(-\pi|y - c|/a) + \underline{C}_1 \exp(-\pi y/a) & ; \ y > 0 \\ \underline{C}_2 \exp(\sqrt{(\pi/a)^2 + \alpha^2} y) & ; \ y < 0. \end{cases}$$

The constants $\underline{C}_1$ and $\underline{C}_2$ follow from the boundary conditions in $y = 0$. In this case the continuity of the $y$-component of the magnetic flux density $\vec{B}$ is equivalent to the continuity of the vector potential:

$$\exp(-\pi c/a) + \underline{C}_1 = \underline{C}_2 .$$

The continuity of the tangential component of the magnetic field requires

$$\frac{1}{\mu_0} \underline{B}_x(x, y > 0)|_{y \to 0} = \frac{1}{\mu} \underline{B}_x(x, y < 0)|_{y \to 0}$$

$$\Leftrightarrow \quad \frac{\mu}{\mu_0} \left[ \frac{\partial \underline{A}(x, y > 0)}{\partial y} \right]_{y \to 0} = \frac{\partial \underline{A}(x, y < 0)}{\partial y} \Big|_{y \to 0}$$

$$\Leftrightarrow \quad \mu/\mu_0 \, \pi/a \, [\exp(-\pi c/a) - \underline{C}_1] = \sqrt{(\pi/a)^2 + \alpha^2} \, \underline{C}_2$$

$$\Rightarrow \quad \underline{C}_2 = \frac{2\pi\mu/\mu_0}{\pi\mu/\mu_0 + \sqrt{\pi^2 + (\alpha a)^2}} \exp(-\pi c/a)$$

$$\Rightarrow \quad \underline{C}_1 = \frac{\pi\mu/\mu_0 - \sqrt{\pi^2 + (\alpha a)^2}}{\pi\mu/\mu_0 + \sqrt{\pi^2 + (\alpha a)^2}} \exp(-\pi c/a) .$$

Hence the magnetic field is determined.

The limit $\alpha \to 0$ or rather $\omega \to 0$ leads to the result

$$\underline{C}_2|_{\omega \to 0} = (1 + k) \exp(-\pi c/a) ; \qquad \underline{C}_1|_{\omega \to 0} = k \exp(-\pi c/a)$$

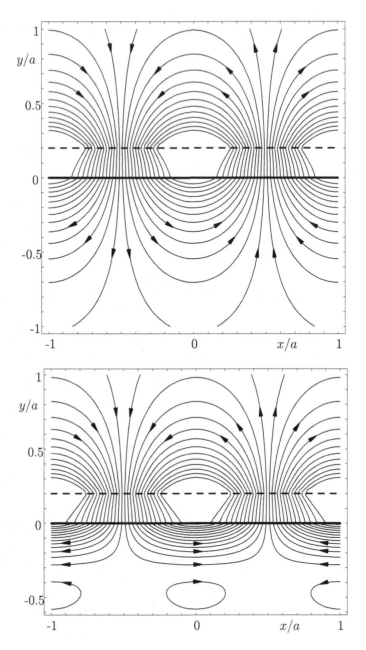

Fig. 5.5–1: Magnetic lines of force for $c/a = 0.2$, $\mu/\mu_0 = 10$, and $\delta/a = 1$ (top) and
$\delta/a = 0.2$ (bottom)

$$k = (\mu - \mu_0)/(\mu + \mu_0) \, .$$

Here we can recognize the method of images for vector potentials of stationary magnetic fields at a permeable half-space, where the exciting vector potential, like in the present case, has no $y$-component:

$$\vec{A}(x,y)\Big|_{\substack{\varphi=0 \\ \omega \to 0}} = \begin{cases} \vec{A}_e(x,y) + k\,\vec{A}_e(x,-y) & ; \quad y \geq 0 \\ (1+k)\,\vec{A}_e(x,y) & ; \quad y \leq 0 \end{cases} \quad ; \quad \vec{e}_y\,\vec{A}_e = 0$$

with $\vec{A}_e(x,y) = \vec{\underline{A}}_e(x,y)\big|_{\varphi=0}$.

At high frequencies it is

$$\underline{C}_1\big|_{\omega \to \infty} = -\exp(-\pi c/a)$$

and the vector potential becomes

$$\vec{\underline{A}}(x,y)\Big|_{\substack{\varphi=0 \\ \omega \to \infty}} = A_e(x,y) - \vec{A}_e(x,-y) \, ; \quad y > 0 \, ; \qquad \vec{e}_y\,\vec{A}_e = 0.$$

This is the method of images for a conducting half-space with high-frequency excitation in $y > 0$, which is different to the method for stationary fields at a high-permeable half-space:

$$\vec{\underline{A}}(x,y)\Big|_{\substack{\varphi=0 \\ \omega \to 0 \\ \mu \to \infty}} = \vec{A}_e(x,y) + \vec{A}_e(x,-y) \, ; \quad y > 0 \, ; \qquad \vec{e}_y\,\vec{A}_e = 0 \, .$$

With a high-frequency excitation the field vanishes in $y \leq 0$ and the plane $y = 0$ carries a current sheet

$$\underline{K}(x) = \int_{-\infty}^{0} \underline{J}(x,y)dy = -\kappa j \omega \int_{-\infty}^{0} A(x,y)dy$$

$$= -\frac{j\omega\kappa\mu_0 a\underline{K}_0}{2\pi} \cos\left(\pi\,\frac{x}{a}\right) \underline{C}_2 \, \frac{1}{\sqrt{(\pi/a)^2 + \alpha^2}}\Bigg|_{\omega \to \infty}$$

$$= -\frac{\mu_0 a\underline{K}_0}{2\pi} \cos\left(\pi\,\frac{x}{a}\right) \frac{1}{\mu}\underline{C}_2 \, \frac{\alpha^2}{\sqrt{(\pi/a)^2 + \alpha^2}}\Bigg|_{\omega \to \infty}$$

$$= -\frac{\mu_0 a\underline{K}_0}{2\pi} \cos\left(\pi\,\frac{x}{a}\right) \frac{1}{\mu}\frac{2\pi\mu/\mu_0}{\alpha a}\exp(-\pi c/a)\,\alpha\Bigg|_{\omega \to \infty}$$

$$= -2\,\frac{1}{\mu_0}\frac{\partial A_e}{\partial y}\Bigg|_{y=0} = 2\left(\vec{e}_y \times \vec{\underline{H}}_e\big|_{y=0}\right)\vec{e}_z \, .$$

Hence the high-frequency current sheet can be calculated directly from the exciting magnetic field.

The time-averaged power loss $\overline{P}_v/l$ in a section $2a$ on the $x$-axis per unit length in $z$-direction is

$$\overline{P}_v/l = -\frac{1}{2} \int_{x=0}^{2a} \text{Re}\left\{\underline{\vec{E}} \times \underline{\vec{H}}^*\right\}_{y=0} \vec{e}_y \, dx$$

$$= -\frac{1}{2} \int_{x=0}^{2a} \text{Re}\left\{\vec{e}_z \, \underline{E}(x, y = 0) \times \vec{e}_x \underline{H}^*(x, y = 0)\right\} \vec{e}_y \, dx$$

$$= -\frac{1}{2} \text{Re}\left\{\int_{x=0}^{2a} \underline{E}(x, 0) \, \underline{H}^*(x, 0) \, dx\right\}$$

$$= -\frac{1}{2} \text{Re}\left\{\int_{x=0}^{2a} -j\omega \underline{A}(x, y = 0) \frac{1}{\mu} \left.\frac{\partial \underline{A}(x, y < 0)}{\partial y}\right|_{y \to 0}^* dx\right\}$$

$$= \frac{1}{2} \text{Re}\left\{\int_{x=0}^{2a} j\omega \frac{\mu_0^2 a^2 |\underline{K}_0|^2}{4\pi^2} \cos^2\left(\pi \frac{x}{a}\right) \frac{|\underline{C}_2|^2}{\mu} \left[\sqrt{(\pi/a)^2 + \alpha^2}\right]^* dx\right\}$$

$$\overline{P}_v/l = -\frac{|\underline{K}_0|^2}{8\pi^2 \kappa} \left(\frac{\mu_0}{\mu}\right)^2 |\underline{C}_2|^2 \text{Re}\left\{(\alpha a)^2 \sqrt{\pi^2 + (\alpha a)^2}\right\}.$$

Finally the equation for the magnetic lines of force is given by

$$d\vec{s} \times \vec{H} = d\vec{s} \times 1/\mu \,\text{rot}\, [\vec{e}_z A(x, y)] = 1/\mu \, d\vec{s} \times (\text{grad}\, A \times \vec{e}_z)$$

$$= 1/\mu \left[ \text{grad}\, A \underbrace{(d\vec{s}\,\vec{e}_z)}_{=0} - \vec{e}_z \underbrace{(d\vec{s}\,\text{grad}\, A)}_{dA} \right] = 0,$$

thus at the time $t = 0$ it is $\quad \text{Re}\{\underline{A}(x, y)\} = \text{const}.$

## 5.6 Induced Current Distribution by a Moving Conductor

A $y$-directed current sheet in $z = 0$ moves with constant velocity $v$ in positive $x$-direction. At the time $t = 0$ it is described by

$$\vec{K}(x, t = 0) = \vec{e}_y \, K_0 \cos(\pi x/a).$$

In $z > 0$ the permeability is $\mu_1$ and the conductivity is $\kappa_1$, whereas in $z < 0$ the permeability is $\mu_2$ and the conductivity is $\kappa_2$.

Find the induced current distribution and analyze the limit of a high velocity $v$.

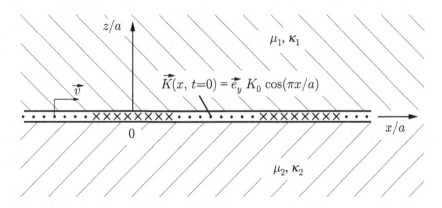

The field is described by a $y$-directed vector potential $\vec{A} = \vec{e}_y A(x, z)$, that satisfies the differential equation

$$\Delta A = \frac{\partial^2 A}{\partial x^2} + \frac{\partial^2 A}{\partial z^2} = \kappa\mu\,\frac{\partial A}{\partial t}\,; \qquad \vec{B} = \mu\vec{H} = \operatorname{rot}\vec{A} = \vec{e}_z\,\frac{\partial A}{\partial x} - \vec{e}_x\,\frac{\partial A}{\partial z}\,.$$

In the moving coordinate system $(\xi = x - vt, z)$ applies

$$\frac{\partial A}{\partial t} = \frac{\partial A}{\partial \xi}\,\frac{\partial \xi}{\partial t} = -v\,\frac{\partial A}{\partial \xi}\,; \qquad \frac{\partial^2 A}{\partial \xi^2} + \frac{\partial^2 A}{\partial z^2} + \kappa\mu v\,\frac{\partial A}{\partial \xi} = 0$$

and for the given excitation the following approach holds

$$A_{1,2}(\xi, z) = \operatorname{Re}\left\{\left\{\begin{array}{c} \underline{C}_1\exp(-\gamma_1 z) \\ \underline{C}_2\exp(\gamma_2 z) \end{array}\right\}\exp(j\,\pi\,\xi/a)\right\}\,; \qquad \begin{array}{c} z > 0 \\ z < 0 \end{array}$$

with

$$-(\pi/a)^2 + \gamma_{1,2}^2 + j\kappa_{1,2}\mu_{1,2}v\pi/a = 0\,; \qquad \gamma_{1,2} = \sqrt{(\pi/a)^2 - j\kappa_{1,2}\mu_{1,2}v\pi/a} =$$

$$= \pi/a\sqrt{1 - j\lambda_{1,2}}\,; \qquad \lambda_{1,2} = \frac{\kappa_{1,2}\mu_{1,2}va}{\pi}\,.$$

The constants $\underline{C}_1$ and $\underline{C}_2$ follow from the evaluation of the boundary conditions in $z = 0$. For the tangential component of the magnetic field it follows

$$\left[-H_\xi(\xi, z < 0) + H_\xi(\xi, z > 0)\right]_{z \to 0} = K(\xi) = K_0\cos(\pi\xi/a)\,.$$

With

$$H_\xi = -\frac{1}{\mu}\,\frac{\partial A}{\partial z}$$

one gets

$$\operatorname{Re}\left\{(\underline{C}_1\gamma_1/\mu_1 + \underline{C}_2\gamma_2/\mu_2)\exp(j\pi\xi/a)\right\} = K_0\cos(\pi\xi/a)\,.$$

The normal component of the magnetic flux density is continuous in $z = 0$ and therefore in this case also the vector potential. Hence it is $\underline{C}_1 = \underline{C}_2 = \underline{C}$.

$$\underline{C} = \frac{K_0}{\gamma_1/\mu_1 + \gamma_2/\mu_2}$$

$$A_{1,2}(x - vt, z) = A_{1,2}(x, z, t) = \mathrm{Re}\{\underline{C} \exp(\mp \gamma_{1,2} z) \exp(j\pi(x - vt)/a)\}$$

The fields are

$$E_{1,2}(x, z, t) = -\frac{\partial}{\partial t} A_{1,2}(x, z, t) =$$

$$= \mathrm{Re}\{j\, \underline{C}\pi\, v/a \, \exp(\mp \gamma_{1,2}\, z)\, \exp(j\,\pi\,\xi/a)\}; \quad \xi = x - vt$$

and

$$\vec{H}_{1,2}(x, z, t) = \frac{1}{\mu_{1,2}} \left[ \vec{e}_z \frac{\partial A_{1,2}}{\partial x} - \vec{e}_x \frac{\partial A_{1,2}}{\partial z} \right] =$$

$$= \frac{1}{\mu_{1,2}} \mathrm{Re}\{\underline{C} \left(\pm\vec{e}_x\, \gamma_{1,2} + \vec{e}_z\, j\,\pi/a\right) \exp(\mp\gamma_{1,2} z) \exp(j\pi\xi/a)\} \, .$$

In the limit of a high velocity the induced currents take the form of current sheets next to the plane $z = 0$. For the lower half-space applies

$$K_2(\xi) = \int_{-\infty}^{0} J_2(x, z, t)|_{v\to\infty} \, dz = \kappa \int_{-\infty}^{0} E_2(x, z, t)|_{v\to\infty} \, dz$$

$$K_2(\xi) = -\kappa \int_{-\infty}^{0} \frac{\partial A_2}{\partial t}\bigg|_{v\to\infty} dz = -\kappa \int_{-\infty}^{0} \frac{\partial A_2}{\partial \xi} \frac{\partial \xi}{\partial t}\bigg|_{v\to\infty} dz$$

$$K_2(\xi) = \mathrm{Re}\left\{ j\, \frac{\kappa_2 v\pi}{a} \frac{K_0}{\gamma_1/\mu_1 + \gamma_2/\mu_2} \exp(j\pi\xi/a) \int_{-\infty}^{0} \exp(\gamma_2 z) dz \right\}\bigg|_{v\to\infty}$$

$$= \mathrm{Re}\left\{ \frac{j v\pi}{a} \frac{K_0 \exp(j\pi\xi/a)}{-j\sqrt{\kappa_1/\kappa_2\, \mu_2/\mu_1}(v\pi/a)^2 - jv\pi/a} \right\}\bigg|_{v\to\infty} =$$

$$K_2(\xi) = -\frac{K_0}{1 + \sqrt{\kappa_1/\kappa_2\, \mu_2/\mu_1}} \cos(\pi\xi/a) \, .$$

The analog result for the upper half-space is

$$K_1(\xi) = -\kappa \int_{0}^{\infty} \frac{\partial A_1}{\partial \xi} \frac{\partial \xi}{\partial t}\bigg|_{v\to\infty} dz = -\frac{K_0}{1 + \sqrt{\kappa_2/\kappa_1\, \mu_1/\mu_2}} \cos(\pi\xi/a) \, .$$

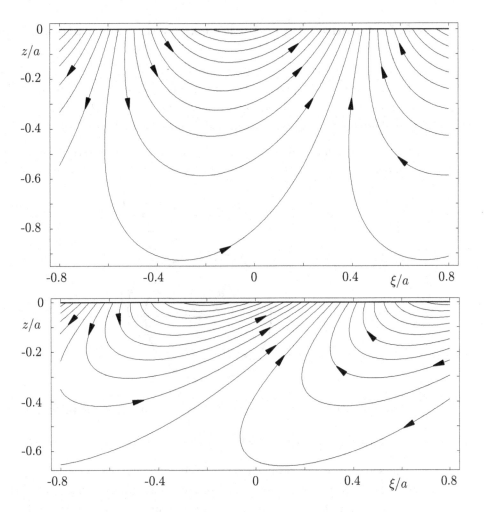

Fig. 5.6–1: Magnetic lines of force for $\lambda_2 = 0.5$ (top) and $\lambda_2 = 3$ (bottom) with $\mu_1 \to \infty$

Finally the equation for the magnetic lines of force at the time $t = 0$ is

$$A_{1,2}(\xi, z) = \mathrm{Re}\{\underline{C}\, \exp(\mp\gamma_{1,2}z + j\pi\xi/a)\} = \mathrm{const}$$

$$A_{1,2}(\xi, z) = \frac{\mu_1 K_0 a}{\pi}\, \mathrm{Re}\left\{\frac{\exp\left(\mp\pi z/a\sqrt{1 - j\lambda_{1,2}}\right)\exp\left(j\pi\xi/a\right)}{\sqrt{1 - j\lambda_1} + \mu_1/\mu_2\sqrt{1 - j\lambda_2}}\right\} = \mathrm{const}$$

with $\lambda_{1,2} = \kappa_{1,2}\,\mu_{1,2}\,v\,a/\pi$; $\quad 0 \le \lambda_{1,2} < \infty$.

## 5.7  Conducting Cylinder Exposed to a Rotating Magnetic Field

A cylinder of conductivity $\kappa$, permeability $\mu$, radius $a$, and infinite length is exposed to a rotating magnetic field with constant magnitude $H_0$. The field is directed perpendicular to the $z$-axis of the cylinder and rotates around it with the angular velocity $\vec{\omega} = \vec{e}_z \omega$. The permeability $\mu_0$ of the surrounding space is constant.

Calculate the induced current distribution and the time-averaged power loss per unit length.

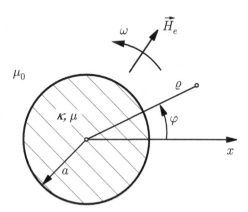

For the exciting rotating field in the homogeneous non-conducting space we can write

$$\vec{H}_e(t) = H_0 \left[ \vec{e}_x \cos(\omega t) + \vec{e}_y \sin(\omega t) \right] = \mathrm{Re} \left\{ (\vec{e}_x - j\,\vec{e}_y)\, H_0 \exp(j\omega t) \right\} .$$

The solution of the problem requires to calculate the effect of only one component $\vec{e}_x H_0 \cos(\omega t) = \vec{e}_x \mathrm{Re} \left\{ \underline{H}_0 \exp(\omega t) \right\}$ of the exciting field. The effect of the second component results from changing the phase and applying a transformation of coordinates.

The exiting vector potential of the subproblem is given by

$$\underline{\vec{A}}_e = \vec{e}_z\, \underline{A}_e(\varrho, \varphi) ; \qquad \underline{\vec{B}}_e = \mathrm{rot}\, \underline{\vec{A}}_e = \mu_0\, \underline{\vec{H}}_e = \vec{e}_x \mu_0\, \underline{H}_0,$$

with

$$\underline{\vec{B}}_e = \vec{e}_\varrho\, \frac{1}{\varrho} \frac{\partial \underline{A}_e}{\partial \varphi} - \vec{e}_\varphi\, \frac{\partial \underline{A}_e}{\partial \varrho} ; \qquad \underline{\vec{H}}_e = (\vec{e}_\varrho \cos\varphi - \vec{e}_\varphi \sin\varphi) \underline{H}_0$$

and thus

$$\underline{A}_e(\varrho, \varphi) = \mu_0\, \underline{H}_0\, \varrho \sin\varphi .$$

It is a solution of the differential equation $\Delta \underline{A}_e(\varrho, \varphi) = \Delta(R(\varrho)\Phi(\varphi)) = 0$ with the general solution functions $R_n(\varrho) = \{\varrho^n, \varrho^{-n}\}$ and $\Phi_n(\varphi) = \{\cos(n\varphi)\,;\,\sin(n\varphi)\}$. Obviously the solution for the exciting field is restricted to the terms with $n = 1$.

Inside the cylinder $\varrho < a$ the resulting vector potential $\vec{\underline{A}} = \vec{e}_z \underline{A}(\varrho, \varphi)$ must satisfy the differential equation

$$\Delta \underline{A} = \frac{\partial^2 \underline{A}}{\partial \varrho^2} + \frac{1}{\varrho} \frac{\partial \underline{A}}{\partial \varrho} + \frac{1}{\varrho^2} \frac{\partial^2 \underline{A}}{\partial \varphi^2} = \alpha^2 \underline{A}; \qquad \alpha^2 = j\omega\kappa\mu = 2j/\delta^2.$$

The solution functions are

$$R_n(\alpha\varrho) = \left\{ \begin{array}{c} I_n(\alpha\varrho) \\ K_n(\alpha\varrho) \end{array} \right\} \quad \text{and} \quad \Phi_n(\varphi).$$

With the given excitation and because of the singularity of the modified Bessel function $K_n$ at $\varrho = 0$ the approach for the resulting vector potential reads with $n = 1$

$$\underline{A}(\varrho, \varphi) = \mu_0 \underline{H}_0 a \left\{ \begin{array}{c} \underline{c} \dfrac{I_1(\alpha\varrho)}{I_1(\alpha a)} \\ \varrho/a + \underline{b}a/\varrho \end{array} \right\} \sin\varphi; \qquad \begin{array}{c} \varrho \leq a \\ \\ \varrho \geq a \end{array}.$$

The evaluation of the boundary conditions on $\varrho = a$ leads to the constants $\underline{c}$ and $\underline{b}$.

The continuity of the normal component of the flux density $\vec{e}_\varrho \vec{\underline{B}}$ and thus in this case the continuity of the vector potential gives

$$\underline{c} = 1 + \underline{b}.$$

A second equation results from the continuity of the tangential component of the magnetic field $\vec{e}_\varrho \times \vec{\underline{H}}$

$$\frac{1}{\mu} \frac{\partial}{\partial \varrho} \underline{A}(\varrho < a, \varphi)|_{\varrho \to a} - \frac{1}{\mu_0} \frac{\partial}{\partial \varrho} \underline{A}(\varrho > a, \varphi)|_{\varrho \to a} = 0$$

that leads to

$$\frac{\mu_0}{\mu} \underline{c} \frac{\alpha a I_1'(\alpha a)}{I_1(\alpha a)} = 1 - \underline{b}; \qquad \underline{c} = \frac{2 I_1(\alpha a)}{I_1(\alpha a)(1 - \mu_0/\mu) + \mu_0/\mu \, \alpha a I_0(\alpha a)}.$$

In the limit $\omega \to 0$ applies

$$I_0(\alpha\varrho)|_{\omega \to 0} = 1; \quad I_1(\alpha\varrho)|_{\omega \to 0} \approx \frac{\alpha\varrho}{2}; \quad \underline{c}|_{\omega \to 0} = (1 + k); \quad \underline{b}|_{\omega \to 0} = \frac{\mu - \mu_0}{\mu + \mu_0} = k$$

$$\underline{A}(\varrho, \varphi)|_{\omega \to 0} = \left\{ \begin{array}{c} (1 + k)\underline{A}_e(\varrho, \varphi) \\ \underline{A}_e(\varrho, \varphi) + k\underline{A}_e(a^2/\varrho, \varphi) \end{array} \right\} \Bigg|_{\omega \to 0}; \qquad \begin{array}{c} \varrho < a \\ \\ \varrho > a \end{array}.$$

This is the method of images for stationary fields at a permeable cylinder.

The induced current distribution $\underline{\vec{J}} = \vec{e}_z \underline{J} = -\vec{e}_z j\omega\kappa\underline{A}$ is

$$\underline{J}(\varrho,\varphi) = -2\frac{\mu_0}{\mu}(\alpha a)^2 \frac{\underline{H}_0}{a} \frac{I_1(\alpha\varrho)}{I_1(\alpha a)(1-\mu_0/\mu)+\mu_0/\mu\,\alpha a I_0(\alpha a)} \sin\varphi\,.$$

It vanishes for $\omega \to 0$ and in the high-frequency limit only a current sheet at $\varrho = a$ exists

$$\underline{K}(\varphi) = \int\limits_0^a \underline{J}(\varrho,\varphi)\big|_{\omega\to\infty} \frac{\varrho}{a}\,d\varrho\,; \qquad \underline{\vec{K}}(\varphi) = 2\vec{e}_\varrho \times \underline{\vec{H}}_e\big|_{\substack{\varrho=a\\\omega\to\infty}} = \vec{e}_z\underline{K}(\varphi)\,,$$

that is related to the magnitude of the magnetic field at $\varrho = a$.

$$\vec{e}_\varphi\underline{\vec{H}}\big|_{\substack{\varrho=a\\\omega\to\infty}} = -\frac{1}{\mu_0}\frac{\partial}{\partial\varrho}\,\underline{A}(\varrho>a,\varphi)\big|_{\substack{\varrho=a\\\omega\to\infty}} =$$

$$= -\underline{H}_0(1-\underline{b})\sin\varphi\big|_{\omega\to\infty} = -2\underline{H}_0\sin\varphi = 2\vec{e}_\varphi\underline{\vec{H}}_e\big|_{\substack{\varrho=a\\\omega\to\infty}}$$

$$\underline{K}(\varphi) = -2\underline{H}_0\sin\varphi = 2\underline{H}_{e\varphi}\big|_{\substack{\varrho=a\\\omega\to\infty}}$$

The time-averaged power loss in the cylinder per unit length is given by

$$\overline{P}_v/l = -\frac{1}{2}\,\mathrm{Re}\left\{\int\limits_0^{2\pi}\underline{\vec{E}}(a)\times\underline{\vec{H}}^*(a)\vec{e}_\varrho a\,d\varphi\right\} =$$

$$= \frac{1}{2}\mathrm{Re}\left\{\int\limits_0^{2\pi}\underline{E}(a)\underline{H}_\varphi^*(a)a\,d\varphi\right\} = \frac{1}{2}\mathrm{Re}\left\{\int\limits_0^{2\pi}j\omega\underline{A}(a)\underline{H}_0^*(1-\underline{b}^*)\sin\varphi a\,d\varphi\right\}$$

$$\overline{P}_v/l = \frac{1}{2}a^2\pi\mu_0\,|\underline{H}_0|^2\,\mathrm{Re}\left\{j\omega(2\underline{c}-|\underline{c}|^2)\right\} = -\mu_0\pi a^2\,|\underline{H}_0|^2\,\omega\,\mathrm{Im}\left\{\underline{c}\right\}\,.$$

The calculation of the total current distribution induced by the rotating field requires the superposition of the field described by the second part of the initial $z$-directed vector potential $\underline{A}_e = -\mu_0\underline{H}_0\,\varrho\cos\varphi$.

This is done by the substitution of $\sin\varphi$ with $\quad\sin\varphi - j(-\cos\varphi) = j\exp(-j\varphi)$ in the expression for the current density.

For the vector potential $\underline{\vec{A}}_D = \vec{e}_z\underline{A}_D$ of the rotating field applies in analogy

$$\underline{A}_D(\varrho,\varphi) = j\mu_0\,\underline{H}_0 a\left\{\begin{array}{l}\underline{c}\,\dfrac{I_1(\alpha\varrho)}{I_1(\alpha a)}\\[2mm]\varrho/a+\underline{b}a/\varrho\end{array}\right\}\exp(-j\varphi)$$

and the field follows from the given relations.

The equation for magnetic lines of force $d\vec{s}\times\vec{H} = 0$ at the time $t = 0$ is $\mathrm{Re}\left\{\underline{A}_D(\varrho,\varphi)\right\} = $ const. Examples of magnetic lines of force are shown in the figures.

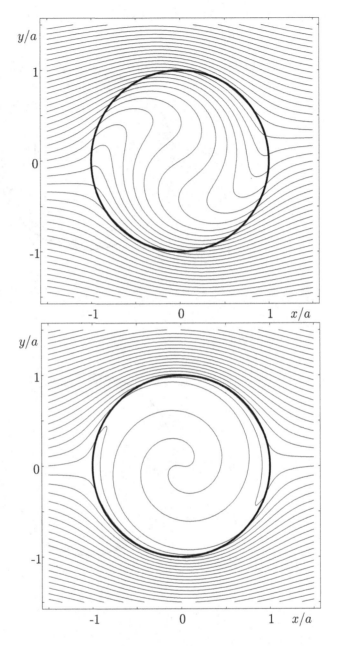

Fig. 5.7–1: Magnetic lines of force in a conducting permeable cylinder, that is exposed to a rotating magnetic field at time $t = 0$

Parameters: $\mu/\mu_0 = 1$, $a/\delta = 3$ (top), $a/\delta = 10$ (bottom)

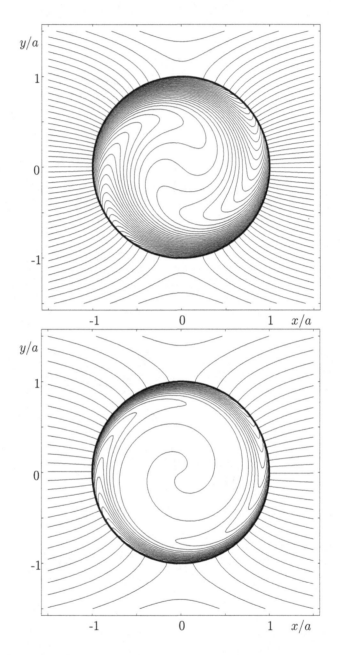

Fig. 5.7–2: Magnetic lines of force in a conducting permeable cylinder, that is exposed
        to a rotating magnetic field at time $t = 0$
        Parameters: $\mu/\mu_0 = 10^3$, $a/\delta = 5$ (top), $a/\delta = 10$ (bottom)

## 5.8 Power Loss and Energy Balance inside a Conducting Sphere Exposed to the Transient Field of a Conductor Loop

A thin conductor loop is positioned at $r = r_E$ and $\vartheta = \vartheta_E$ in front of a sphere with radius $a < r_E$, conductivity $\kappa$, and permeability $\mu$. The permeability of the surrounding space is $\mu_0$.

Find the induced current distribution in the sphere, when in contrast to problem 5.3 the exciting current $i(t)$ has an arbitrary time-dependence. Additionally calculate the time-averaged power loss in case of a time-harmonic excitation and analyze the energy balance when a direct current is turned off.

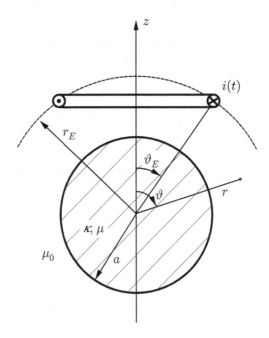

In this case the rotationally symmetric field is described by a second order vector potential

$$\vec{W} = \vec{r}\, W(r, \vartheta, t) \; ; \qquad \frac{\partial W}{\partial \varphi} = 0.$$

With it, the field is given by

$$\vec{A} = \operatorname{rot} \vec{W} = -\vec{r} \times \operatorname{grad} W \; ; \quad \vec{B} = \operatorname{rot} \vec{A} = \operatorname{rot} \operatorname{rot} \vec{W} \; ; \quad \vec{J} = -\kappa \frac{\partial \vec{A}}{\partial t} = \kappa \vec{E} \,.$$

## Time-Harmonic Field

In the case of a time-harmonic excitation the current is described by a complex amplitude $\underline{i}_0$. The exciting field of the conductor loop in the homogeneous space of permittivity $\mu_0$ follows from the exciting potential $\vec{\underline{W}}_E$.

$$\vec{\underline{W}}_E(r,\vartheta) = \vec{r}\,\underline{W}_E\,; \qquad \Delta\underline{W}_E = 0\,; \qquad r \neq r_E$$

$$\underline{W}_E(r,\vartheta) = \sum_{n=0}^{\infty} \left\{ \begin{array}{l} \underline{E}_n\,(r/r_E)^n \\ \underline{D}_n\,(r_E/r)^{n+1} \end{array} \right\} P_n(\cos\vartheta)\,; \qquad \begin{array}{l} r < r_E \\ r > r_E \end{array}$$

$$\vec{\underline{A}}_E = -\vec{r} \times \operatorname{grad}\underline{W}_E\,; \qquad \vec{\underline{B}}_E = \operatorname{rot}\operatorname{rot}\vec{\underline{W}}_E = \operatorname{grad}\left[\frac{\partial}{\partial r}(r\underline{W}_E)\right] = \mu_0\vec{\underline{H}}_E$$

For the determination of the constants $\underline{D}_n$ and $\underline{E}_n$ we evaluate the boundary conditions on the sphere $r = r_E$. The continuity of the normal component of the magnetic flux density requires

$$\left.\frac{\partial^2}{\partial r^2}(r\underline{W}_E)\right|_{\substack{r<r_E \\ r\to r_E}} = \left.\frac{\partial^2}{\partial r^2}(r\underline{W}_E)\right|_{\substack{r>r_E \\ r\to r_E}}\,; \qquad \underline{D}_n = \underline{E}_n\,.$$

Moreover the tangential component of the magnetic field yields

$$\underline{H}_{E\vartheta}\Big|_{\substack{r>r_E \\ r\to r_E}} - \underline{H}_{E\vartheta}\Big|_{\substack{r<r_E \\ r\to r_E}} = \underline{K}(\vartheta)\,,$$

where $\underline{K}(\vartheta)$ is a $\varphi$-directed current sheet, that is zero except for $\vartheta = \vartheta_E$.

With

$$\underline{H}_{E\vartheta} = \frac{1}{\mu_0}\frac{1}{r}\frac{\partial^2}{\partial r\partial\vartheta}(r\underline{W}_E)$$

and $u = \cos\vartheta$ it follows

$$\underline{H}_{E\vartheta} = -\frac{1}{\mu_0}\frac{1}{r}\sum_{n=0}^{\infty}\underline{E}_n\left\{ \begin{array}{l} (n+1)(r/r_E)^n \\ -n(r_E/r)^{n+1} \end{array} \right\}\sqrt{1-u^2}\,P_n'(u)\,; \qquad \begin{array}{l} r < r_E \\ r > r_E \end{array}$$

and thus it is

$$\frac{1}{\mu_0}\sum_{n=0}^{\infty}(2n+1)\underline{E}_n\sqrt{1-u^2}\,P_n'(u) = r_E\underline{K}(u)\,.$$

Now the multiplication with $\sqrt{1-u^2}P_k'(u)$ and integration from $u = -1$ to $u = +1$

leads to

$$\underline{E}_n = \frac{\mu_0}{2n(n+1)} \int\limits_{u=-1}^{+1} \underline{K}(u) r_E \sqrt{1-u^2}\, P_n'(u)\, du$$

$$= \frac{\mu_0}{2n(n+1)} P_n'(u_E)(1-u_E^2) \underbrace{\int\limits_{\vartheta=0}^{\pi} \underline{K}(\vartheta) r_E\, d\vartheta}_{i_0}\,; \qquad u_E = \cos\vartheta_E$$

$$\underline{E}_n = \frac{1}{2}\mu_0\, i_0 (1-u_E^2)\frac{P_n'(u_E)}{n(n+1)} = \frac{1}{2}\mu_0\, i_0 \frac{P_{n-1}(u_E) - P_{n+1}(u_E)}{2n+1}$$

where the orthogonality relation for Legendre polynomials

$$\int\limits_{-1}^{+1}(1-u^2)P_n'(u)\,P_k'(u)\,du = \begin{cases} \dfrac{2n(n+1)}{2n+1} & ;\quad k = n \\ 0 & ;\quad k \neq n \end{cases}$$

has been applied.

In the presence of the conducting sphere the field is described by the second order potential

$$\vec{\underline{W}} = \vec{r}\underline{W}\,; \qquad \Delta\underline{W} = \alpha^2\underline{W}$$

$$\underline{W} = \sum_{n=0}^{\infty} \left\{ \begin{array}{l} \underline{A}_n(a/r)^{n+1} + \underline{E}_n\left\{\begin{array}{l}(r/r_E)^n \\ (r_E/r)^{n+1}\end{array}\right\} \\ \underline{B}_n R_n(\alpha r) \end{array}\right\} P_n(\cos\vartheta)\,; \qquad \begin{array}{l} r > a \\[2ex] r < a \end{array}$$

with the spherical Bessel functions

$$R_n(\alpha r) = \sqrt{\frac{\pi}{2\alpha r}}\, I_{n+1/2}(\alpha r)\,; \qquad \alpha^2 = j\omega\kappa\mu$$

that satisfy the recurrence relations

$$\alpha r\,[R_{n-1} - R_{n+1}] = (2n+1)R_n\,; \qquad \alpha r\, R_n'(\alpha r) = nR_n + \alpha r\, R_{n+1}$$

$$(\alpha r)^2\, R_n''(\alpha r) = \left[n(n-1) + (\alpha r)^2\right]R_n - 2\alpha r\, R_{n+1}\,.$$

In $r < a$ the magnetic flux density is

$$\vec{\underline{B}} = \mathrm{grad}\,\frac{\partial}{\partial r}(r\underline{W}) - \vec{r}\alpha^2\underline{W}$$

and because of the continuity of the normal component $\vec{e}_r \underline{\vec{B}}$ in $r = a$ and thus the continuity of $\underline{W}$ it is

$$\underline{A}_n + \underline{E}_n (a/r_E)^n = \underline{B}_n R_n(\alpha a).$$

A second equation results from the continuity of the tangential component of $\underline{\vec{H}} = \underline{\vec{B}}/\mu$

$$\frac{\mu_0}{\mu} \underline{B}_n \left[ (n+1) R_n(\alpha a) + \alpha a \, R_{n+1}(\alpha a) \right] = -n \underline{A}_n + (n+1) \underline{E}_n (a/r_E)^n.$$

This system of equations has the solution

$$\underline{B}_n = \frac{2n+1}{n R_n(\alpha a) + \mu_0/\mu \left[ (n+1) R_n(\alpha a) + \alpha a \, R_{n+1}(\alpha a) \right]} \left( \frac{a}{r_E} \right)^n \underline{E}_n$$

$$\underline{A}_n = \left[ \frac{(2n+1) R_n(\alpha a)}{n R_n(\alpha a) + \mu_0/\mu \left[ (n+1) R_n(\alpha a) + \alpha a \, R_{n+1}(\alpha a) \right]} - 1 \right] \left( \frac{a}{r_E} \right)^n \underline{E}_n.$$

The induced current distribution in the sphere is

$$\underline{\vec{J}} = \frac{1}{\mu} \operatorname{rot} \underline{\vec{B}} = -\frac{\alpha^2}{\mu} \operatorname{rot} (\vec{r}\underline{W}) = \frac{\alpha^2}{\mu} \vec{r} \times \operatorname{grad} \underline{W} = \frac{\alpha^2}{\mu} \vec{e}_\varphi \frac{\partial \underline{W}}{\partial \vartheta} = \vec{e}_\varphi \underline{J}$$

$$\underline{J}(r, \vartheta) = -\frac{\alpha^2}{\mu} \sum_{n=0}^{\infty} \underline{B}_n R_n(\alpha r) \sqrt{1 - u^2} \, P_n'(u).$$

For the determination of the time-averaged power loss in the sphere we integrate the real part of the complex Poynting vector over the surface $a_K$ of the sphere.

$$\overline{P}_v = -\frac{1}{2} \oint_{a_K} \operatorname{Re}\left\{ \underline{\vec{E}} \times \underline{\vec{H}}^* \right\} d\vec{a}_K$$

$$= -\frac{1}{2} \operatorname{Re} \left\{ \oint_{a_K} \left[ \vec{e}_\varphi \underline{E}_\varphi \times (\vec{e}_r \underline{H}_r^* + \vec{e}_\vartheta \underline{H}_\vartheta^*) \right] \vec{e}_r \, a^2 \sin \vartheta d\vartheta d\varphi \right\}$$

$$= \pi a^2 \operatorname{Re} \left\{ \int_{-1}^{+1} \underline{E}_\varphi(a, u) \, \underline{H}_\vartheta^*(a, u) du \right\}$$

$$\underline{E}_\varphi = -j\omega \sum_{n=0}^{\infty} \underline{B}_n R_n(\alpha r) \sqrt{1 - u^2} \, P_n'(u)$$

$$\underline{H}_\vartheta = -\frac{1}{\mu r} \sum_{n=0}^{\infty} \underline{B}_n \left[ (n+1) R_n(\alpha r) + \alpha r R_{n+1}(\alpha r) \right] \sqrt{1 - u^2} P_n'(u)$$

Finally the solution is

$$\overline{P}_v = \frac{-2\pi \omega a}{\mu} \sum_{n=0}^{\infty} \frac{n(n+1)}{2n+1} |\underline{B}_n|^2 \operatorname{Im}\left\{ R_n \left[ (n+1) R_n(\alpha a) + \alpha a \, R_{n+1}(\alpha a) \right] \right\}.$$

**Transient Field**

The basis for the calculation of the field in case of an arbitrary time-dependence of the exciting current is the field, that emerges when a constant current is turned off. Therefore we need at first knowledge of the stationary field, that is excited by a constant current $I_0$ in the conductor loop. The stationary field follows from the corresponding field equations, or alternatively from the limit $\omega \to 0$ in the time-periodic case.

$$\underline{B}_n R_n(\alpha r)\big|_{\omega \to 0} = \frac{2n+1}{n+\mu_0/\mu(n+1)} \left(\frac{r}{r_E}\right)^n E_n = B_n^{(0)} \left(\frac{r}{a}\right)^n$$

$$B_n^{(0)} = \frac{2n+1}{n+\mu/\mu_0(n+1)} \left(\frac{a}{r_E}\right)^n E_n$$

$$E_n = \mu_0 I_0 \frac{1-u_E^2}{2} \frac{P_n'(u_E)}{n(n+1)}; \qquad \underline{A}_n\big|_{\omega \to 0} = \left[\frac{2n+1}{n+\mu_0/\mu(n+1)} - 1\right] \left(\frac{a}{r_E}\right)^n E_n$$

$$\underline{A}_n\big|_{\omega \to 0} = \frac{(n+1)(1-\mu_0/\mu)}{n+(n+1)\mu_0/\mu} \left(\frac{a}{r_E}\right)^n E_n = A_n^{(0)}$$

After turning off the current $I_0$ at the time $t = 0$ the second order potential $\vec{W}_A = \vec{r} W_A(r, \vartheta, t)$ is given by

$$W_A = \sum_{n=0}^{\infty} \sum_{s=1}^{\infty} \left\{ \begin{matrix} a_{ns}\, j_n(\lambda_{ns} r) \\ b_{ns}\, j_n(\lambda_{ns} a)(a/r)^{n+1} \end{matrix} \right\} P_n(\cos\vartheta) \exp(-t/\tau_{ns}); \qquad \tau_{ns} = \frac{\kappa\mu}{\lambda_{ns}^2}.$$

It is a solution of the differential equation

$$\Delta W_A = \left\{ \begin{matrix} \kappa\mu \dfrac{\partial W_A}{\partial t} & ; & r < a \\ 0 & ; & r > a \end{matrix} \right.$$

with the spherical Bessel functions

$$j_n(\lambda_{ns} r) = \sqrt{\frac{\pi}{2\lambda_{ns} r}} J_{n+1/2}(\lambda_{ns} r).$$

The vector potential and the magnetic flux density are

$$\vec{A}_A = \operatorname{rot}(\vec{r} W_A); \qquad \vec{B}_A = \operatorname{rot}\operatorname{rot}(\vec{r} W_A) = \operatorname{grad}\frac{\partial}{\partial r}(r W_A) - \kappa\mu \frac{\partial \vec{W}_A}{\partial t}.$$

For the determination of the constants $a_{ns}$ and $b_{ns}$ we evaluate the boundary conditions in $r = a$ and furthermore at the time $t = 0$ the field matches the initial stationary solution. The continuity of the normal component of the magnetic flux density requires

$$\left( r \frac{\partial^2}{\partial r^2} (rW_A) - \kappa \mu r^2 \frac{\partial W_A}{\partial t} \right)\Bigg|_{\substack{r<a \\ r \to a}} = r \frac{\partial^2}{\partial r^2} (rW_A)\Bigg|_{\substack{r>a \\ r \to a}} .$$

With the recurrence relations and the derivatives for spherical Bessel functions this results in

$$a_{ns} = b_{ns} .$$

Now with this result the second continuity relation for the tangential component of the magnetic field results in the eigenvalue equation

$$\lambda_{ns} a\, j_{n-1}(\lambda_{ns} a) - n\, j_n(\lambda_{ns} a)\,(1 - \mu/\mu_0) = 0$$

and thus leads to the eigenvalues $\lambda_{ns} = x_{ns}/a$. Hence the potential is

$$W_A = \sum_{n=0}^{\infty} \sum_{s=1}^{\infty} a_{ns} \left\{ \begin{array}{l} j_n(x_{ns}r/a) \\ j_n(x_{ns})(a/r)^{n+1} \end{array} \right\} P_n(u) \exp(-t/\tau_{ns}); \qquad \tau_{ns} = \frac{\kappa \mu a^2}{x_{ns}^2} .$$

Finally the constant $a_{ns}$ follows from the initial stationary field at the time $t = 0$.

$$\sum_{n=0}^{\infty} \sum_{s=1}^{\infty} a_{ns}\, j_n(x_{ns}r/a)\, P_n(u) = \sum_{n=0}^{\infty} B_n^{(0)} (r/a)^n\, P_n(u)$$

$$B_n^{(0)} (r/a)^n = \sum_{s=1}^{\infty} a_{ns} j_n(x_{ns}r/a); \qquad u = \cos \vartheta$$

Due to the orthogonality relation for the the spherical Bessel functions the multiplication with $j_n(x_{np}r/a)w(r)$ and $w(r) = 2\lambda_{np}r^2/\pi$ and integration results in

$$a_{ns} = \frac{2 j_{n+1}(x_{ns})}{x_{ns}[j_n^2(x_{ns}) - j_{n-1}(x_{ns})j_{n+1}(x_{ns})]}\, B_n^{(0)} .$$

With it, the transient field for $t > 0$ is determined. In case of an arbitrary time-dependence of the current $i(t)$ for $t > 0$ and $i(t) = 0$ for $t < 0$ the potential $W_A(\vec{r}, t)$ is given, if the exponential function $\exp(-t/\tau_{ns})$ in $W_A$ is replaced by the function

$$f_{ns}(t) = \frac{1}{I_0} \left[ i(0)\,[1 - \exp(-t/\tau_{ns})] + \int_0^t \frac{di(t - \tau)}{dt}\,[1 - \exp(-\tau/\tau_{ns})]\, d\tau \right] .$$

## Energy Balance

When the initially constant current is turned off the energy balance requires

$$W_{m0} - W_{mA} - W_{ms} = 0 \, .$$

Here $W_{m0}$ is the energy of the magnetic field inside the sphere before the turn-off, $W_{mA}$ is the energy transferred trough the surface $r = a$ after the turn-off, and $W_{ms}$ is the dissipated energy inside the sphere for $t > 0$. It follows

$$\frac{1}{2}\mu \int\limits_{v_K} H_A^2(\vec{r}, t < 0) dv - \int\limits_{t=0}^{\infty} \oint\limits_{a_K} \vec{E}_A(\vec{r}, t) \times \vec{H}_A(\vec{r}, t)\, d\vec{a}_K\, dt = W_{m0} - W_{mA} =$$

$$= \int\limits_{t=0}^{\infty} \int\limits_{v_K} \vec{E}_A(\vec{r}, t)\, \vec{J}_A(\vec{r}, t) dv dt = W_{ms} \, .$$

The energy of the stationary magnetic field in $r < a$ is derived by

$$W_{m0} = \frac{1}{2\mu} \int\limits_{v_K} B_A^2(\vec{r}, t < 0) dv = \frac{1}{2\mu} \int\limits_{v_K} [\text{rot rot}\,(\vec{r} W_A(\vec{r}, t = 0))]^2\, dv$$

$$= \frac{1}{2\mu} \int\limits_{v_K} \text{grad}^2 \left[\frac{\partial}{\partial r}(rW_A)\right]_{t=0} dv = \frac{1}{2\mu} \oint\limits_{a_K} \frac{\partial}{\partial r}(rW_A)\, \frac{\partial^2}{\partial r^2}(rW_A)\Big|_{\substack{r=a \\ t=0}} da_K$$

$$\Rightarrow \quad W_{m0} = \frac{2\pi a}{\mu} \sum\limits_{n=0}^{\infty} \frac{n(n+1)^2}{2n+1} \left(B_n^{(0)}\right)^2 \, .$$

Here Green's second identity has been applied.

The integration of the Poynting vector leads to

$$W_{mA} = \int\limits_{t=0}^{\infty} \oint\limits_{a_K} \vec{E}_A \times \vec{H}_A d\vec{a}_K\, dt \, ; \qquad d\vec{a}_K = \vec{e}_r da_K \, .$$

With

$$\vec{H}_A \times \vec{E}_A = \frac{1}{\mu} \left[\text{grad}\, \frac{\partial}{\partial r}(rW_A) - \kappa\mu\vec{r}\, \frac{\partial W_A}{\partial t}\right] \times \left[\vec{r} \times \text{grad}\, \frac{\partial W_A}{\partial t}\right] = -\vec{S}$$

$$\frac{\vec{r}}{r}\vec{S} = -\frac{r}{\mu} \text{grad}_s \left[\frac{\partial}{\partial r}(rW_A)\right] \text{grad}_s \frac{\partial W_A}{\partial t} \, ; \qquad \text{grad}_s = \frac{\vec{e}_\vartheta}{r} \frac{\partial}{\partial \vartheta}$$

$$= -\frac{1}{\mu} \frac{\partial}{\partial \vartheta} \left[\frac{\partial}{\partial r}(rW_A)\right] \frac{1}{r} \frac{\partial}{\partial \vartheta} \frac{\partial W_A}{\partial t}$$

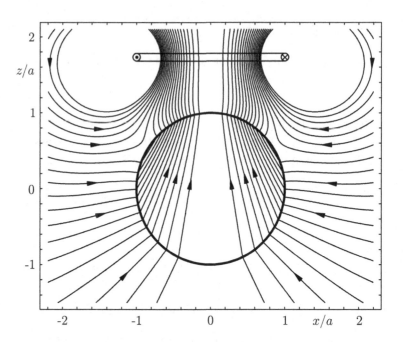

Fig. 5.8–1: Magnetic lines of force before the turn-off with $\mu = 100\mu_0$,
$\kappa = 2 \cdot 10^6$ [S/m], $r_E/a = 2$, and $\vartheta_E = \pi/6$

it follows:

$$
W_{mA} = -\frac{1}{\mu} \int\limits_{t=0}^{\infty} \int\limits_{0}^{2\pi} \int\limits_{-1}^{+1} \left[ \sum_{n=0}^{\infty} \sum_{s=1}^{\infty} a_{ns} \left[ x_{ns}\, j_{n-1}(x_{ns}) - n\, j_n(x_{ns}) \right] \cdot \right.
$$

$$
\left. \cdot \left[ -\sqrt{1-u^2}\, P_n'(u) \right] \exp(-t/\tau_{ns}) \right] \frac{1}{a} \left[ \sum_{k=0}^{\infty} \sum_{r=1}^{\infty} a_{kr}\, j_k(x_{kr}) \left[ -\sqrt{1-u^2}\, P_k'(u) \right] \cdot \right.
$$

$$
\left. \cdot \frac{-x_{kr}^2}{\kappa\mu a^2}\, \exp(-t/\tau_{kr}) \right] \cdot a^2 (-du)\, d\varphi dt
$$

$$
= -\frac{2\pi a}{\mu} \int\limits_{t=0}^{\infty} \sum_{n=0}^{\infty} \left[ \sum_{s=1}^{\infty} a_{ns}\, (x_{ns} j_{n-1}(x_{ns}) - n j_n(x_{ns})) \right] \left[ \sum_{r=1}^{\infty} a_{nr}\, j_n(x_{nr}) \right] \cdot
$$

$$
\cdot \int\limits_{-1}^{+1} (1-u^2) P_n'^2(u) du\, \frac{-x_{nr}^2}{\kappa\mu a^2}\, \exp(-t/\tau_{ns})\, \exp(-t/\tau_{nr})\, dt
$$

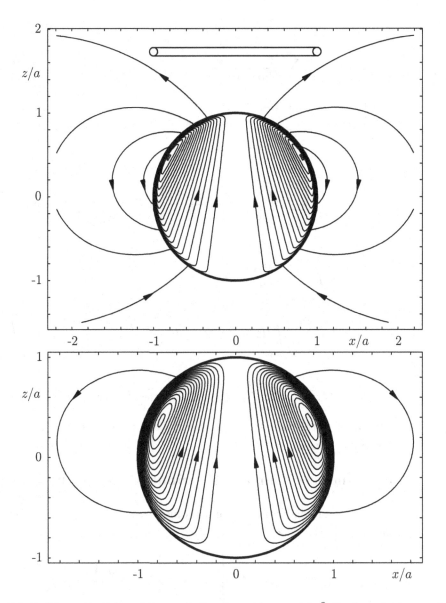

Fig. 5.8–2: Magnetic lines of force at the time $t/\tau_{01} = 10^{-3}$ (top) and $t/\tau_{01} = 10^{-2}$ (bottom) after the turn-off

Parameters see Fig. 5.8–1.

$$W_{mA} = \frac{2\pi a}{\mu} \sum_{n=0}^{\infty} \frac{2n(n+1)}{2n+1} \sum_{s=1}^{\infty} \sum_{r=1}^{\infty} a_{ns}\, a_{nr}\, j_n(x_{nr}) \cdot$$

$$\cdot \left[ x_{ns} j_{n-1}(x_{ns}) - n j_n(x_{ns}) \right] \cdot \frac{x_{nr}^2}{\kappa \mu a^2} \int_{t=0}^{\infty} \exp(-t/\tau_{ns}) \exp(-t/\tau_{nr})\, dt$$

$$W_{mA} = \frac{4\pi a}{\mu} \sum_{n=0}^{\infty} \frac{n(n+1)}{2n+1} \sum_{s=1}^{\infty} \sum_{r=1}^{\infty} a_{ns}\, a_{nr}\, j_n(x_{nr})\, x_{nr}^2\, \frac{[x_{ns} j_{n-1}(x_{ns}) - n j_n(x_{ns})]}{x_{ns}^2 + x_{nr}^2}.$$

Finally the last integral gives

$$W_{ms} = \kappa \int_{t=0}^{\infty} \int_{v_K} E_A^2(\vec{r}, t)\, dv\, dt =$$

$$= \kappa \int_{t=0}^{\infty} \int_{v_K} \left[\vec{r} \times \mathrm{grad}\, \frac{\partial W_A}{\partial t}\right]^2 dv\, dt = \kappa \int_{t=0}^{\infty} \int_{v_K} \left(r\, \mathrm{grad}_s \frac{\partial W_A}{\partial t}\right)^2 dv\, dt$$

$$W_{ms} = \kappa \int_{t=0}^{\infty} \int_{v_K} \left[\frac{\partial^2 W_A}{\partial \vartheta \partial t}\right]^2 dv\, dt$$

$$= 2\pi\kappa \sum_{n=0}^{\infty} \int_{u=-1}^{+1} (1 - u^2) P_n'^2(u)\, du \sum_{s=1}^{\infty} \int_{r=0}^{a} a_{ns}^2 j_n^2\left(x_{ns}\frac{r}{a}\right) r^2\, dr \int_{t=0}^{\infty} \frac{\exp(-2t/\tau_{ns})}{\tau_{ns}^2}\, dt$$

$$W_{ms} = \frac{\pi a}{\mu} \sum_{n=0}^{\infty} \frac{n(n+1)}{2n+1} \sum_{s=1}^{\infty} a_{ns}^2\, x_{ns}^2 \left[j_n^2(x_{ns}) - j_{n-1}(x_{ns}) j_{n+1}(x_{ns})\right].$$

The proof of the validity of the energy balance is done by a numerical calculation of
the expressions above.

## 5.9   Induced Current Distribution in a Conducting Cylinder

A cylinder of conductivity $\kappa$ and radius $a$ is concentrically surrounded by an axially
directed current sheet

$$\vec{K}(\varphi, t) = \vec{e}_z\, \mathrm{Re} \left\{ \begin{Bmatrix} \underline{K}_0 \\ -\underline{K}_0 \end{Bmatrix} \exp(j\omega t) \right\}; \qquad \begin{matrix} 0 < \varphi < \pi \\ \pi < \varphi < 2\pi \end{matrix}$$

on the cylinder $\varrho = b > a$. The permeability $\mu$ is constant.

Find the induced current distribution and the time-averaged power loss.

The magnetic field is described by an axially directed vector potential
$\vec{A}(\varrho, \varphi) = \vec{e}_z \underline{A}(\varrho, \varphi)$, that satisfies the differential equation

$$\Delta\underline{A} = \frac{1}{\varrho}\frac{\partial}{\partial\varrho}\left(\varrho\frac{\partial\underline{A}}{\partial\varrho}\right) + \frac{1}{\varrho^2}\frac{\partial^2\underline{A}}{\partial\varphi^2} = \begin{cases} \alpha^2\underline{A} & ; \ \varrho < a \\ 0 & ; \ \varrho > a; \varrho \neq b \end{cases} ; \quad \alpha^2 = j\omega\kappa\mu = \frac{2j}{\delta^2}.$$

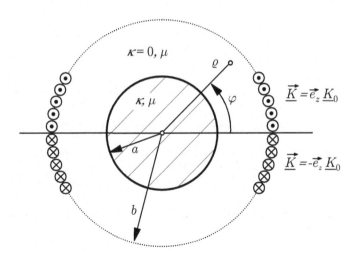

Solution functions are the modified Bessel functions $I_n(\alpha\varrho)$ and $K_n(\alpha\varrho)$ in $\varrho < a$, the powers of the distance to the axis $\varrho^n$ and $\varrho^{-n}$ in $\varrho > a$, and the trigonometric functions $\sin(n\varphi)$ and $\cos(n\varphi)$.

At first the exciting vector potential $\vec{\underline{A}}_e(\varrho,\varphi) = \vec{e}_z\,\underline{A}_e(\varrho,\varphi)$ of the current sheet at $\varrho = b$ in the homogeneous space of permeability $\mu$ will be determined.

$$\Delta\underline{A}_e = 0\,; \quad \varrho \neq b\,; \qquad \underline{A}_e(\varrho,\varphi) = \sum_{n=1}^{\infty}\underline{a}_n\left\{\begin{array}{l}(\varrho/b)^n \\ (b/\varrho)^n\end{array}\right\}\sin(n\varphi)\,; \qquad \begin{array}{l}\varrho \leq b \\ \varrho \geq b\end{array}$$

The continuity of the normal component of the magnetic flux density $\vec{B}$ is identical to the continuity of the vector potential itself and is already satisfied. Here the solution with $n = 0$ is omitted because of

$$\int_0^{2\pi}\underline{K}(\varphi)\varrho\,d\varphi = 0\,.$$

The magnetic flux density $\vec{\underline{B}}_e = \operatorname{rot}\vec{\underline{A}}_e$ is given by

$$\vec{\underline{B}}_e = \operatorname{rot}[\vec{e}_z\underline{A}_e] = \operatorname{grad}\underline{A}_e \times \vec{e}_z = \vec{e}_\varrho\,\frac{1}{\varrho}\,\frac{\partial\underline{A}_e}{\partial\varphi} - \vec{e}_\varphi\,\frac{\partial\underline{A}_e}{\partial\varrho} = \mu\left[\vec{e}_\varrho\underline{H}_{e\varrho} + \vec{e}_\varphi\underline{H}_{e\varphi}\right].$$

For the determination of $\underline{a}_n$ the boundary condition for the tangential component of the magnetic field in $\varrho = b$ is evaluated.

$$\underline{H}_{e\varphi}\Big|_{\substack{\varrho>b \\ \varrho\to b}} - \underline{H}_{e\varphi}\Big|_{\substack{\varrho<b \\ \varrho\to b}} = \underline{K}(\varphi) = \left\{\begin{array}{ll}\underline{K}_0 & ;\quad 0 < \varphi < \pi \\ -\underline{K}_0 & ;\quad \pi < \varphi < 2\pi\end{array}\right.$$

$$\left.\frac{\partial \underline{A}_e}{\partial \varrho}\right|_{\substack{\varrho<b \\ \varrho\to b}} - \left.\frac{\partial \underline{A}_e}{\partial \varrho}\right|_{\substack{\varrho>b \\ \varrho\to b}} = \mu \underline{K}(\varphi)$$

$$\sum_{n=1}^{\infty} 2n\, \underline{a}_n/b \, \sin(n\varphi) = \mu \begin{cases} \underline{K}_0 & ; \quad 0 < \varphi < \pi \\ -\underline{K}_0 & ; \quad \pi < \varphi < 2\pi \end{cases}$$

Now applying the orthogonality relation for trigonometric functions leads to

$$2\pi n\, \underline{a}_n/b = 2\mu \underline{K}_0 \int_{\varphi=0}^{\pi} \sin(n\varphi)d\varphi; \qquad \underline{a}_n = \frac{2\mu \underline{K}_0 b}{\pi n^2}; \qquad n = 2k+1.$$

After inserting the conducting cylinder the ansatz for the resulting vector potential $\underline{A}(\varrho,\varphi)$ is

$$\underline{A}(\varrho,\varphi) = \frac{2\mu \underline{K}_0 b}{\pi} \sum_{k=0}^{\infty} \left\{ \begin{array}{l} \underline{b}_n \, I_n(\alpha\varrho)/I_n(\alpha a) \\ \left\{ \begin{array}{l} (\varrho/b)^n \\ (b/\varrho)^n \end{array} \right\} + \underline{d}_n(a/\varrho)^n \end{array} \right\} \frac{\sin(n\varphi)}{n^2}; \quad \begin{array}{l} \varrho \le a \\ a \le \varrho \le b \\ \varrho \ge b. \end{array}$$

Again the continuity of the normal component of the magnetic flux density $\vec{B}$ in $\varrho = a$ is identical to the continuity of the vector potential.

$$\underline{b}_n = (a/b)^n + \underline{d}_n$$

The continuity of the tangential component $\underline{H}_\varphi = -1/\mu\, \partial\underline{A}/\partial\varrho$ results in

$$\underline{b}_n \, \alpha a \frac{I_n'(\alpha a)}{I_n(\alpha a)} = n\left(\frac{a}{b}\right)^n - n\underline{d}_n \qquad \Rightarrow \qquad \underline{b}_n = 2\left(\frac{a}{b}\right)^n \left[1 + \frac{\alpha a}{n}\frac{I_n'(\alpha a)}{I_n(\alpha a)}\right]^{-1}.$$

Therewith the current density $\underline{\vec{J}}$ inside the cylinder $\varrho < a$ is determined.

$$\underline{\vec{J}} = -j\omega\kappa\underline{\vec{A}} = -\vec{e}_z j\omega\kappa\underline{A}(\varrho,\varphi) = -\vec{e}_z \frac{2}{\pi}(\alpha b)^2 \frac{\underline{K}_0}{b} \sum_{k=0}^{\infty} \underline{b}_n \frac{I_n(\alpha\varrho)}{I_n(\alpha a)} \frac{\sin(n\varphi)}{n^2}; \quad n = 2k+1$$

The time-averaged power loss per unit length $l$ in the cylinder follows from the integration of the Poynting vector over the surface of the cylinder.

$$\overline{P}_v/l = -\frac{1}{2}\int_{\varphi=0}^{2\pi} \text{Re}\left\{\underline{\vec{E}} \times \underline{\vec{H}}^*\right\}\Big|_{\varrho=a} \vec{e}_\varrho ds = \frac{1}{2}\int_{\varphi=0}^{2\pi} \text{Re}\left\{\frac{j\omega\underline{A}}{\mu}\left(\frac{\partial\underline{A}}{\partial\varrho}\right)^*\right\}\Big|_{\varrho=a} a\, d\varphi$$

$$= \frac{\omega a}{2\mu}\left(\frac{2\mu b|\underline{K}_0|}{\pi}\right)^2 \sum_{k=0}^{\infty} \text{Re}\left\{j|\underline{b}_n|^2 \left(\frac{I_n'(\alpha a)}{I_n(\alpha a)}\alpha\right)^* \pi\right\} \frac{1}{n^4}$$

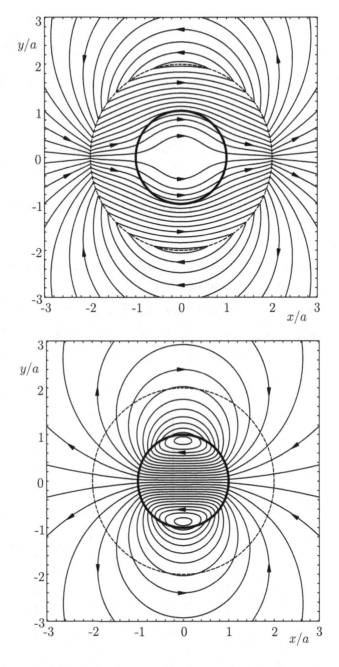

Fig. 5.9–1: Magnetic lines of force at the time $t = 0$ (top) and
$\omega t = \pi/2$ (bottom) with $\varphi_0 = 0$, $\underline{K}_0 = K_0 \exp(j\varphi_0)$,
and $\delta/a = 0.5$

$$\overline{P}_v/l = -\frac{2}{\pi}\omega\mu\left(|\underline{K}_0|\,b\right)^2\sum_{k=0}^{\infty}|\underline{b}_n|^2/n^4\,\mathrm{Im}\left\{\left(\alpha a\,\frac{I_n'(\alpha a)}{I_n(\alpha a)}\right)^*\right\};\quad n = 2k+1$$

Finally the equation for the magnetic lines of force $d\vec{s}\times\underline{\vec{H}} = 0$ at the time $t = 0$ reads

$$d\vec{s}\times\underline{\vec{H}} = d\vec{s}\times\frac{1}{\mu}\mathrm{rot}\,[\vec{e}_z\underline{A}(\varrho,\varphi)] = \frac{1}{\mu}d\vec{s}\times(\mathrm{grad}\,\underline{A}\times\vec{e}_z) =$$

$$= -\vec{e}_z\frac{1}{\mu}d\underline{A} = 0;\quad \mathrm{Re}\{\underline{A}(\varrho,\varphi)\} = \mathrm{const}\ .$$

## 5.10  Cylinder with Stepped Down Diameter

In the plane $z = 0$ the radius of a homogeneous cylinder with conductivity $\kappa$ is stepped down from $b$ in $z < 0$ to $a < b$ in $z > 0$. The conductor carries the current $i(t) =$ $= \mathrm{Re}\,\{\underline{i}_0\exp(j\omega t)\}$ in axial direction.

Calculate the current distribution inside the conductor. The permeability $\mu$ is constant. See also chapters 3 and 4 for the static case $\omega \to 0$.

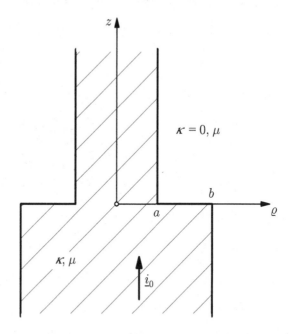

Because of the rotational symmetry the magnetic field is $\varphi$-directed

$$\underline{\vec{H}} = \vec{e}_\varphi\,\underline{H}(\varrho,z)$$

and satisfies the differential equation

$$\text{rot rot}\,\vec{H} + \kappa\mu\frac{\partial\vec{H}}{\partial t} = 0 ; \qquad \text{rot rot}\,\vec{H}(\varrho, z) + \alpha^2\vec{H}(\varrho, z) = 0$$

$$\text{rot}\left[-\vec{e}_\varrho\frac{\partial H}{\partial z} + \vec{e}_z\frac{1}{\varrho}\frac{\partial}{\partial\varrho}(\varrho H)\right] = \vec{e}_\varphi\left[-\frac{\partial^2 H}{\partial z^2} - \frac{\partial}{\partial\varrho}\left(\frac{1}{\varrho}\frac{\partial}{\partial\varrho}(\varrho H)\right)\right] = -\alpha^2\vec{e}_\varphi\,H$$

$$\Rightarrow \quad \frac{\partial^2 H}{\partial\varrho^2} + \frac{1}{\varrho}\frac{\partial H}{\partial\varrho} - \frac{1}{\varrho^2}H + \frac{\partial^2 H}{\partial z^2} = \alpha^2 H .$$

The ansatz

$$\underline{H}(\varrho, z) = \underline{R}(\varrho)\,\underline{Z}(z)$$

leads to the equation

$$\frac{1}{\underline{R}}\left(\frac{d^2\underline{R}}{d\varrho^2} + \frac{1}{\varrho}\frac{d\underline{R}}{d\varrho} - \frac{\underline{R}}{\varrho^2}\right) + \frac{1}{\underline{Z}}\frac{d^2\underline{Z}}{dz^2} = \alpha^2 = j\omega\kappa\mu = 2j/\delta^2$$

that separates to

$$\frac{d^2\underline{R}}{d\varrho^2} + \frac{1}{\varrho}\frac{d\underline{R}}{d\varrho} - \frac{\underline{R}}{\varrho^2} + m^2\underline{R} = 0 ; \qquad \delta = \sqrt{\frac{2}{\omega\kappa\mu}}$$

and

$$\frac{d^2\underline{Z}_m}{dz^2} - (\alpha^2 + m^2)\underline{Z}_m = 0.$$

Solutions are the Bessel functions of order one

$$\underline{R}_m(\varrho) = \underline{C}_1 J_1(m\varrho) + \underline{C}_2 N_1(m\varrho)$$

and the exponential function

$$\underline{Z}_m(z) = \underline{C}\exp\left(\pm\sqrt{\alpha^2 + m^2}\,z\right) .$$

If the dependence on the $z$-coordinate is missing (homogeneous conductor of infinite length), then the differential equation reads

$$\frac{d^2\underline{R}}{d\varrho^2} + \frac{1}{\varrho}\frac{d\underline{R}}{d\varrho} - \frac{\underline{R}}{\varrho^2} - \alpha^2\underline{R} = 0$$

and the solutions are the modified Bessel Functions $I_1(\alpha\varrho)$ and $K_1(\alpha\varrho)$.

The combined approach for the magnetic field of the present problem is

$$\underline{H}(\varrho, z) = \underline{H}_0\begin{cases}\dfrac{I_1(\alpha\varrho)}{I_1(\alpha a)} + \sum\limits_{m1}\underline{C}_{m1}\,J_1(m_1\varrho)\exp(-\sqrt{\alpha^2 + m_1^2}\,z) \; ; & \begin{array}{l}z \geq 0\\ \varrho \leq a\end{array}\\[3ex] \dfrac{a}{b}\dfrac{I_1(\alpha\varrho)}{I_1(\alpha b)} + \dfrac{a}{b}\sum\limits_{m2}\underline{C}_{m2}\,J_1(m_2\varrho)\exp(\sqrt{\alpha^2 + m_2^2}\,z) \; ; & \begin{array}{l}z \leq 0\\ \varrho \leq b\end{array}\end{cases}$$

with $\underline{H}_0 = \underline{i}_0/(2\pi a)$. The preceding terms with modified Bessel functions describe the homogeneous field far away from the discontinuity in $z = 0$.

The unknown parameters $m_1, m_2$ follow from the boundary conditions

$$\vec{e}_\varrho \underline{\vec{J}}\Big|_{\varrho=a,b} = 0 \, ; \quad \underline{\vec{J}} = \vec{e}_\varrho \underline{J}_\varrho + \vec{e}_z \underline{J}_z \, ; \quad \underline{\vec{J}} = \mathrm{rot}\,\underline{\vec{H}} \, ; \quad \underline{J}_\varrho = -\frac{\partial \underline{H}}{\partial z}$$

$$\frac{\partial \underline{H}}{\partial z}\bigg|_{\substack{z>0 \\ \varrho=a}} = \frac{\partial \underline{H}}{\partial z}\bigg|_{\substack{z<0 \\ \varrho=b}} = 0 \quad \Rightarrow \quad \underline{R}_m(\varrho = a)|_{z>0} = \underline{R}_m(\varrho = b)|_{z<0} = 0 \, .$$

With the zeros $x_{1r}$ of the Bessel functions $J_1(x_{1r}) = 0$ it follows

$$J_1(m_1 a) = 0 \quad \Rightarrow \quad m_1 = x_{1r}/a \, ; \quad r = 1, 2, 3, \ldots$$
$$J_1(m_2 b) = 0 \quad \Rightarrow \quad m_2 = x_{1r}/b \, ; \quad r = 1, 2, 3, \ldots \, .$$

Now the magnetic field is

$$\underline{H}(\varrho, z) = \underline{H}_0 \begin{cases} \dfrac{I_1(\alpha\varrho)}{I_1(\alpha a)} + \displaystyle\sum_{r=1}^{\infty} \underline{C}_{1r} J_1(x_{1r}\varrho/a) \, \exp(-\sqrt{\alpha^2 + (x_{1r}/a)^2}\,z) \\[4mm] \dfrac{a}{b}\dfrac{I_1(\alpha\varrho)}{I_1(\alpha b)} + \dfrac{a}{b} \displaystyle\sum_{r=1}^{\infty} \underline{C}_{2r} J_1(x_{1r}\varrho/b) \, \exp(\sqrt{\alpha^2 + (x_{1r}/b)^2}\,z) \end{cases}$$

in the respective range of validity.

Outside the conductor the magnetic field is $\underline{H}(\varrho) = \underline{H}_0\, a/\varrho$.

Finally the constants $\underline{C}_{1r}$ and $\underline{C}_{2r}$ follow from the boundary conditions in $z = 0$.

$$\underline{H}(\varrho, z)\big|_{\substack{z<0 \\ z\to 0}} = \begin{cases} \underline{H}(\varrho, z)|_{\substack{z>0 \\ z\to 0}} & ; \quad \varrho \leq a \\[2mm] \underline{H}_0 a/\varrho & ; \quad \varrho > a \end{cases}$$

$$\frac{a}{b}\frac{I_1(\alpha\varrho)}{I_1(\alpha b)} + \frac{a}{b}\sum_{r=1}^{\infty} \underline{C}_{2r} J_1(x_{1r}\varrho/b) = \begin{cases} \dfrac{I_1(\alpha\varrho)}{I_1(\alpha a)} + \displaystyle\sum_{r=1}^{\infty} \underline{C}_{1r} J_1(x_{1r}\varrho/a) & ; \quad \varrho \leq a \\[4mm] a/\varrho & ; \quad a \leq \varrho \leq b \end{cases}$$

With the expansion

$$\frac{I_1(\alpha\varrho)}{I_1(\alpha a)} = -2\sum_{r=1}^{\infty} \frac{x_{1r}}{x_{1r}^2 + (\alpha a)^2} \frac{J_1(x_{1r}\varrho/a)}{J_0(x_{1r})}$$

it is

$$\frac{a}{b}\sum_{r=1}^{\infty}\left[\underline{C}_{2r}-\frac{2x_{1r}}{x_{1r}^2+(\alpha b)^2}\frac{1}{J_0(x_{1r})}\right]J_1(x_{1r}\varrho/b)=$$

$$=\begin{cases}\displaystyle\sum_{r=1}^{\infty}\left[\underline{C}_{1r}-\frac{2x_{1r}}{x_{1r}^2+(\alpha a)^2}\frac{1}{J_0(x_{1r})}\right]J_1(x_{1r}\varrho/a)&;\quad \varrho<a\\[4mm] a/\varrho&;\quad a<\varrho<b.\end{cases}$$

The multiplication with $J_1(x_{1s}\varrho/b)\varrho d\varrho$, $s=1,2,3,\ldots$, and integration from $\varrho=0$ to $\varrho=b$ gives because of the orthogonality of Bessel functions

$$\frac{a}{b}\left[\underline{C}_{2s}-\frac{2x_{1s}}{(x_{1s}^2+(\alpha b)^2)J_0}\right]\frac{b^2}{2}J_0^2(x_{1s})=\sum_{r=1}^{\infty}\left[\underline{C}_{1r}-\frac{2x_{1r}}{(x_{1r}^2+(\alpha a)^2)J_0(x_{1r})}\right].$$

$$\cdot\underbrace{\int_0^a J_1(x_{1r}\varrho/a)J_1(x_{1s}\varrho/b)\varrho d\varrho}_{\dfrac{x_{1r}J_1(x_{1s}a/b)J_0(x_{1r})}{(x_{1s}/b)^2-(x_{1r}/a)^2}}+\underbrace{\int_a^b a/\varrho\, J_1(x_{1s}\varrho/b)\varrho d\varrho}_{\dfrac{ab}{x_{1s}}[J_0(x_{1s}a/b)-J_0(x_{1s})]}$$

Therewith, the following system of equations results

$$1/2\left[\underline{C}_{2s}-\frac{2x_{1s}}{x_{1s}^2+(\alpha b)^2}\frac{1}{J_0(x_{1s})}\right]J_0^2(x_{1s})-[J_0(x_{1s}a/b)-J_0(x_{1s})]/x_{1s}=$$

$$=J_1(x_{1s}a/b)b/a\sum_{r=1}^{\infty}\left[\underline{C}_{1r}-\frac{2x_{1r}}{x_{1r}^2+(\alpha a)^2}\frac{1}{J_0(x_{1r})}\right]\frac{x_{1r}J_0(x_{1r})}{x_{1s}^2-(x_{1r}b/a)^2}.$$

A second system results from the continuity of the tangential component $\underline{J}_\varrho$ of the current density

$$\underline{J}_\varrho=-\frac{\partial H}{\partial z}\;;\qquad \left.\frac{\partial H}{\partial z}\right|_{\substack{z<0\\z\to0}}=\left.\frac{\partial H}{\partial z}\right|_{\substack{z>0\\z\to0}}\;;\quad \varrho<a$$

$$-\sum_{r=1}^{\infty}\underline{C}_{1r}J_1(x_{1r}\frac{\varrho}{a})\sqrt{\alpha^2+(x_{1r}/a)^2}=\frac{a}{b}\sum_{r=1}^{\infty}\underline{C}_{2r}J_1(x_{1r}\frac{\varrho}{b})\sqrt{\alpha^2+(x_{1r}/b)^2}.$$

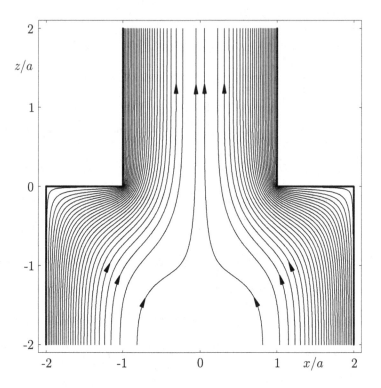

Fig. 5.10–1: Streamlines at the time $\omega t = 0$ with $b/a = 2$ and $b/\delta = 3$

The Multiplication with $J_1(x_{1s}\varrho/a)\varrho d\varrho$ and integration from $\varrho = 0$ to $\varrho = a$ yields

$$- \underline{C}_{1s} 1/2 a^2 J_0^2(x_{1s}) \sqrt{\alpha^2 + (x_{1s}/a)^2} =$$

$$= a/b \sum_{r=1}^{\infty} \underline{C}_{2r} \sqrt{\alpha^2 + (x_{1r}/b)^2} \underbrace{\int_0^a J_1(x_{1r}\varrho/b) J_1(x_{1s}\varrho/a)\varrho d\varrho}_{\dfrac{x_{1s} J_1(x_{1r}a/b) J_0(x_{1s})}{(x_{1r}/b)^2 - (x_{1s}/a)^2}} .$$

Thus the second system of equations reads

$$\underline{C}_{1s} \sqrt{(\alpha a)^2 + x_{1s}^2} J_0(x_{1s}) = 2x_{1s} \sum_{r=1}^{\infty} \underline{C}_{2r} \sqrt{(\alpha b)^2 + x_{1r}^2} \frac{J_1(x_{1r}a/b)}{(x_{1s}b/a)^2 - x_{1r}^2} .$$

The combination of both systems leads to a linear system of equations for the determination of the unknown constants. Exemplary results for the streamlines at different times are presented in the figures.

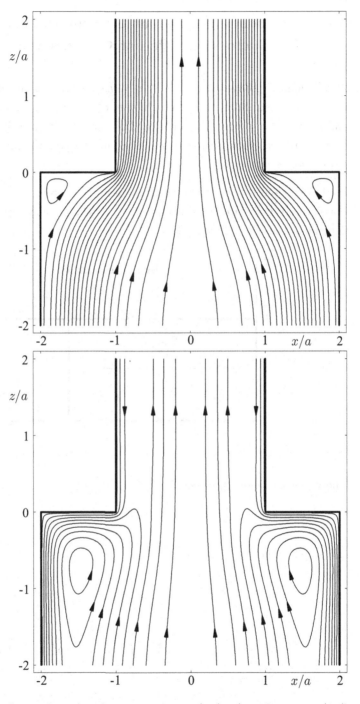

Fig. 5.10–2: Streamlines at the time $\omega t = \pi/4$ (top) and $\omega t = \pi/2$ (bottom) with $b/a = 2$ and $b/\delta = 3$

## 5.11 Frequency-Dependent Current Distribution in Conductors of Different Conductivity

A rectangular groove ($x > 0\,; 0 \le y \le h$) inside a nonconducting material of high permeability is filled with two conductors of conductivity $\kappa_1$ in $x < a$ (conductor (1)) and $\kappa_2$ in $a < x < b$ (conductor (2)). The permittivity $\mu$ inside the groove is constant. In $x = a$ the conductors are electrically connected and the total current carried in axial $z$-direction is $i(t) = \mathrm{Re}\{i_0 \exp(j\omega t)\}$.

Find the magnetic field inside the groove and the current distribution in the conductors. Furthermore calculate the ratio $\eta = i_2/i_1$ of the currents in conductor (1) and (2) in dependence on the frequency $\omega$.

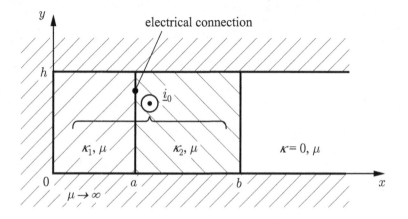

The magnetic field is $y$-directed

$$\vec{\underline{H}}(x) = \vec{e}_y\, \underline{H}(x)$$

and satisfies the differential equation

$$\mathrm{rot}\,\mathrm{rot}\,\vec{\underline{H}} = -j\omega\kappa\mu\vec{\underline{H}}\,; \qquad \frac{d^2\underline{H}}{dx^2} = \alpha_{1,2}^2\underline{H}$$

with

$$\alpha_1^2 = j\omega\kappa_1\mu\,; \qquad \alpha_2^2 = j\omega\kappa_2\mu = 2j/\delta^2 \quad \Rightarrow \quad \alpha_2 = \frac{1+j}{\delta}.$$

The current density is given by

$$\vec{\underline{J}}(x) = \mathrm{rot}\left[\vec{\underline{H}}(x)\right] = \vec{e}_z\,\underline{J}(x) = \vec{e}_z\frac{d\underline{H}(x)}{dx}$$

An ansatz for the magnetic field is

$$\underline{H}(x) = \underline{H}_0 \begin{cases} \underline{C}_1 \dfrac{\sinh(\alpha_1 x)}{\sinh(\alpha_1 a)} & ; \quad 0 \leq x \leq a \\[3mm] \underline{C}_2 \dfrac{\sinh(\alpha_2(x-a))}{\sinh(\alpha_2(b-a))} + \underline{C}_3 \dfrac{\cosh(\alpha_2(x-a))}{\cosh(\alpha_2(b-a))} & ; \quad a \leq x \leq b \\[3mm] 1 & ; \quad x \geq b \end{cases}$$

with $\underline{H}_0 = \underline{i}_0/h$. The evaluation of the boundary conditions in $x = a$ and $x = b$ gives rise to the three constants. From the continuity of the magnetic field in $x = a$ we deduce

$$\underline{C}_1 = \underline{C}_3/\cosh(\alpha_2(b-a)).$$

In $x = b$ applies $\underline{H}(x = b) = \underline{H}_0$ and thus

$$\underline{C}_2 + \underline{C}_3 = 1.$$

Finally the continuity of the electric field in $x = a$ leads to a third equation

$$\vec{\underline{J}} = \operatorname{rot}\vec{\underline{H}}; \qquad 1/\kappa_1 \, d\underline{H}/dx\big|_{\substack{x<a \\ x \to a}} = 1/\kappa_2 \, d\underline{H}/dx\big|_{\substack{x>a \\ x \to a}}$$

$$1/\kappa_1 \, \alpha_1 \, \underline{C}_1 \coth(\alpha_1 a) = 1/\kappa_2 \, \alpha_2 \, \underline{C}_2/\sinh(\alpha_2(b-a)).$$

The solution of this system of equations is

$$\underline{C}_1 = \frac{1}{\cosh(\alpha_2(b-a))(1+\underline{N})}; \qquad \underline{C}_2 = \underline{N}/(1+\underline{N}); \qquad \underline{C}_3 = 1/(\underline{N}+1)$$

$$\underline{N} = \sqrt{\kappa_2/\kappa_1} \coth(\alpha_1 a) \tanh(\alpha_2(b-a)).$$

The current $\underline{i}_1$ in the conductor (1) is

$$\underline{i}_1 = \underline{H}(a) \cdot h; \qquad \underline{i}_1/\underline{i}_0 = \underline{C}_1$$

and for the current $\underline{i}_2$ in the conductor (2) applies

$$\underline{i}_2 = \underline{H}(b) \cdot h - \underline{i}_1; \qquad \underline{i}_2/\underline{i}_0 = \underline{C}_2 + \underline{C}_3 - \underline{C}_1$$

$$\underline{i}_1/\underline{i}_0 = \frac{1}{\cosh(\alpha_2(b-a))(1+\underline{N})}; \qquad \underline{i}_2/\underline{i}_0 = \left[1 - \frac{1}{\cosh(\alpha_2(b-a))(1+\underline{N})}\right].$$

Hence the current ratio $\eta$ is

$$\underline{\eta} = \frac{\underline{i}_2}{\underline{i}_1} = (\underline{N}+1)\cosh(\alpha_2(b-a)) - 1.$$

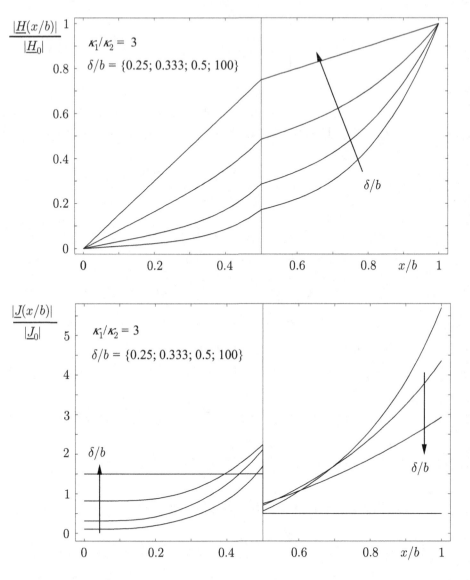

Fig. 5.11–1: Magnitude of the magnetic field (top) and current density (bottom) for different normalized skin depths $\delta/b$ and $b/a = 2$ ($\underline{J}_0 = \underline{H}_0/b$)

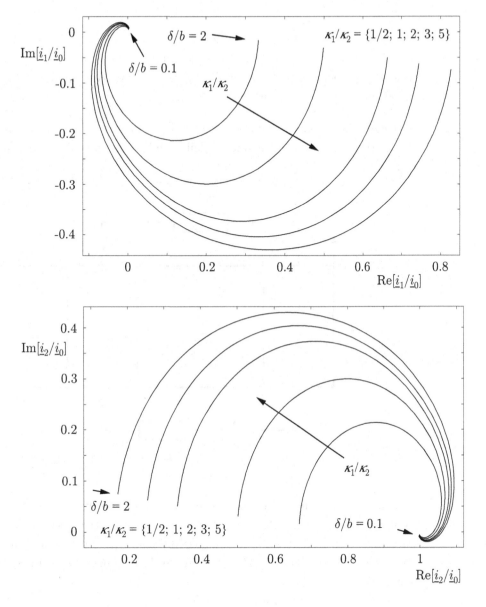

Fig. 5.11–2: Parametric plot of the normalized currents $\underline{i}_1/\underline{i}_0$ (top) and $\underline{i}_2/\underline{i}_0$ (bottom) in dependence on the normalized skin depth $\delta/b$ for different conductivity ratios $\kappa_1/\kappa_2$

## 5.12    Electric Circuit with Massive Conductors

The groove ($x > 0 \, ; 0 \leq y \leq h$) inside a high-permeable material is filled in $x < a$ and $a < b < x < c$ with material of conductivity $\kappa$. The conductors are of length $l$ and connect a generator with alternating voltage of complex amplitude $\underline{U}_0$ and internal impedance $\underline{R}_G$ in $z = 0$ with a terminating impedance $\underline{R}_E$ in $z = l \gg c$.

Calculate the current distribution inside the conductors. The influence of the junctions in $z = 0$ and $z = l$ on the field is negligible and the permeability $\mu$ inside the groove is constant.

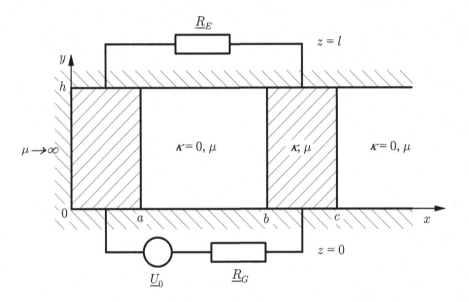

We suppose that the conductor $x < a$ carries a current with complex amplitude $\underline{i}$ in positive $z$-direction and the conductor $b < x < c$ carries the reverse current in negative direction. Because of the boundary conditions in $y = 0$ and $y = h$ the magnetic field is $y$-directed $\underline{\vec{H}}(x) = \vec{e}_y \underline{H}(x)$ and inside the conductors it satisfies the differential equation

$$\frac{d^2 \underline{H}}{dx^2} = \alpha^2 \underline{H} \qquad \text{with} \qquad \alpha^2 = j\omega\kappa\mu .$$

Furthermore the magnetic field vanishes in $x = 0$ and $x = c$, because of the high permeable boundary and because the total current vanishes. Thus the integral of the magnetic field in $x = c$ from $y = 0$ and $y = h$ has to be zero. Hence the magnetic field is given by

$$\underline{H}(x) = \frac{\underline{i}}{h} \left\{ \begin{array}{ll} \dfrac{\sinh(\alpha x)}{\sinh(\alpha a)} & ; \quad 0 \le x \le a \\[2ex] \dfrac{\sinh(\alpha(x-c))}{\sinh(\alpha(b-c))} & ; \quad b \le x \le c \end{array} \right. .$$

In the space $a \le x \le b$ between the conductors the magnetic field is homogeneous $\vec{\underline{H}}(x) = \vec{e}_y \underline{i}/h$.

Finally for the determination of the amplitude $\underline{i}$ the law of induction has to be analyzed.

$$\oint\limits_C \vec{\underline{E}} d\vec{s} = -\frac{\partial}{\partial t} \int\limits_a \vec{\underline{B}} d\vec{a}$$

A suitable contour for the integration runs through the inner boundary $x = a, b$ of the conductors. This results in

$$(\underline{E}(a) - \underline{E}(b)) l + \underline{i} (\underline{R}_E + \underline{R}_G) - \underline{U}_0 = -j\omega\mu \underline{H}(a)(b-a) l.$$

The electric field is given by

$$\vec{\underline{E}}(x) = 1/\kappa \operatorname{rot} [\vec{e}_y \underline{H}(x)] = 1/\kappa \vec{e}_z \frac{d\underline{H}(x)}{dx} = \vec{e}_z \underline{E}(x)$$

$$\underline{E}(x) = \frac{\underline{i}}{\kappa a h} \alpha a \left\{ \begin{array}{ll} \dfrac{\cosh(\alpha x)}{\sinh(\alpha a)} & ; \quad 0 \le x \le a \\[2ex] \dfrac{\cosh(\alpha(x-c))}{\sinh(\alpha(b-c))} & ; \quad b \le x \le c \end{array} \right. .$$

It follows

$$\frac{\underline{i} l}{\kappa a h} \alpha a \left[ \coth(\alpha a) - \coth(\alpha(b-c)) \right] + \underline{i} (\underline{R}_E + \underline{R}_G) + j\omega\mu(b-a) \frac{\underline{i}}{h} l = \underline{U}_0$$

and for the complex amplitude of the current applies

$$\underline{i} = \underline{U}_0 \left[ \underline{R}_E + \underline{R}_G + j\omega\mu(b-a)l/h + \alpha/\kappa \left[\coth(\alpha a) - \coth(\alpha(b-c))\right] l/h \right]^{-1}$$

$$= \underline{U}_0 \left[ \underline{R}_E + \underline{R}_G + j\omega L_A + R(\omega) + j\omega L_i(\omega) \right]^{-1} .$$

The substitution $L_A$ is the outer self-inductance

$$L_A = \mu(b-a)l/h$$

and $\underline{Z}$ is the inner impedance

$$\underline{Z}(\omega) = R(\omega) + j\omega L_i(\omega) = \alpha/\kappa \left[\coth(\alpha a) - \coth(\alpha(b-c))\right] l/h$$

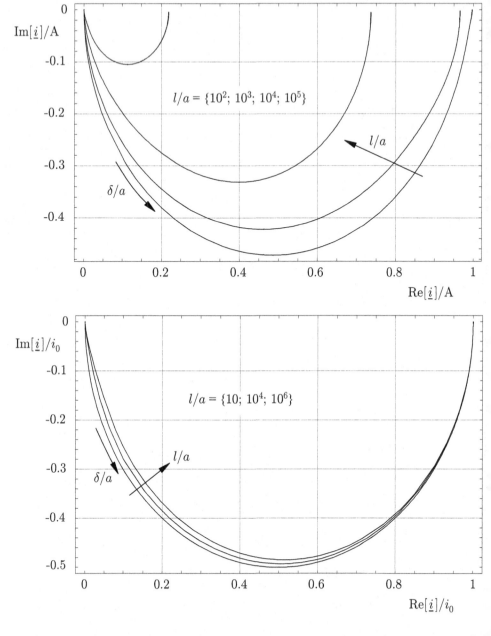

Fig. 5.12–1: Parametric plot of the current with $h = c - b = a$, $b/a = 2$, $\underline{U}_0 = 1\,[\text{V}]$
$i_0 = \underline{i}|_{\omega \to 0}$, and $\underline{R}_G = 0$
$R_E = 1\,[\Omega]$ (top) and $\underline{R}_E = j\omega L$ with $L = 10^{-5}\,[\text{H}]$ (bottom)

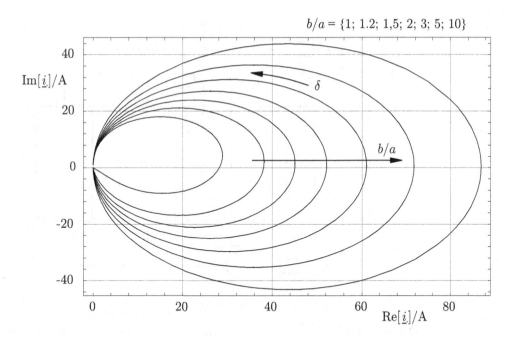

Fig. 5.12–2: See caption of Fig. 5.12–1, but $\underline{R}_E = 1/(j\omega C)$ and $C = 10^{-5}$ [F]

with a frequency-dependent resistance $R(\omega)$ and the reactance defines the inner self-inductance $L_i(\omega)$.

In the limit $\omega = 0$ the current is continuous and described by

$$i = \frac{U_0}{R_E + R_G + l(1/a + 1/(c - b))/(\kappa h)} \, ,$$

where we have assumed that the complex amplitudes $\underline{R}_E, \underline{R}_G$, and $\underline{U}_0$ switch over to the real valued amplitudes.


## 5.13   Magnetically Coupled System of Conductors

The space $0 < y < h$ of permeability $\mu$ is surrounded by high-permeable material $\mu \to \infty$ in $y \le 0$ and $y \ge h$. Two pairs of conductors with conductivity $\kappa$ are located in this space bounded by the planes $x = a$ and $x = b, x = -a$ and $x = -b$, $x = c$ and $x = d$, and $x = -c$ and $x = -d$. The inner pair is short-circuited in $z = 0$ and in $z = l \gg d$ terminated by the impedance $\underline{R}_E$. In the outer pair a current is injected at $z = 0$ by a generator with voltage $\underline{U}_0$ and internal impedance $\underline{R}_G$ and in $z = l$ the outer conductor pair is terminated by the impedance $\underline{R}_S$.

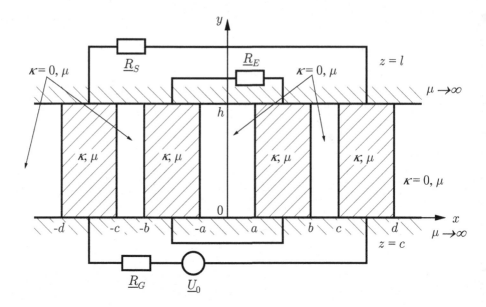

Calculate the currents in the conductors, when the influence of the junctions in $z = 0$ and $z = l$ on the field is negligible.

We suppose that the conductors $a \leq |x| \leq b$ and $c \leq |x| \leq d$ carry the currents $\underline{i}_2$ and $\underline{i}_1$. Because of the high-permeable boundary and the geometry of the conductors the magnetic field has only a $y$-component and depends just on the $x$-coordinate $\underline{\vec{H}} = \vec{e}_y \underline{H}(x)$. Inside the conductors it satisfies the differential equation

$$\frac{d^2 \underline{H}}{dx^2} = \alpha^2 \underline{H} \, ; \qquad \alpha^2 = j\omega\kappa\mu = \frac{2j}{\delta^2} \quad \Rightarrow \quad \alpha = \frac{1+j}{\delta} \, .$$

As the magnetic field vanishes in $|x| > d$ and is homogeneous between the conductors the ansatz for $\underline{H}(x)$ is

$$\underline{H}(x) = \begin{cases} \dfrac{\underline{i}_1}{h} \dfrac{\sinh(\alpha(|x| - d))}{\sinh(\alpha(c - d))} & ; \quad c \leq |x| \leq d \\[2ex] \dfrac{\underline{i}_1}{h} & ; \quad b \leq |x| \leq c \\[2ex] \dfrac{(\underline{i}_1 + \underline{i}_2)}{h} \dfrac{\sinh(\alpha(|x| - b))}{\sinh(\alpha(a - b))} + \dfrac{\underline{i}_1}{h} \dfrac{\sinh(\alpha(|x| - a))}{\sinh(\alpha(b - a))} & ; \quad a \leq |x| \leq b \\[2ex] \dfrac{(\underline{i}_1 + \underline{i}_2)}{h} & ; \quad |x| \leq a \, . \end{cases}$$

This approach already satisfies the continuity relations of the magnetic field.

The homogeneous field between the conductors is defined by the currents $\underline{i}_1$ and $\underline{i}_2$ which are unknown so far. For the calculation of the currents we evaluate the law o induction

$$\oint_C \vec{E} d\vec{s} = -\frac{\partial}{\partial t} \int_a \vec{B} d\vec{a} \, ; \qquad \underline{E} = \frac{1}{\kappa} \mathrm{rot}\, \underline{\vec{H}} = \vec{e}_z \frac{1}{\kappa} \frac{d\underline{H}}{dx} = \vec{e}_z \underline{E}(x)$$

along the boundaries of the conductors. It is convenient to integrate along the inner contour of both conductor pairs. The integration over $x = \pm c$ leads to

$$\underline{E}(-c)l - \underline{E}(c)l + \underline{i}_1(\underline{R}_G + \underline{R}_S) - \underline{U}_0 =$$

$$= -j\omega\mu \left[ 2\underline{i}_1/h(c-b)l + 2(\underline{i}_1 + \underline{i}_2)/hal + 2l \int_a^b \underline{H}(x)dx \right]$$

and a second integration over $x = \pm a$ gives

$$\underline{E}(-a)l - \underline{E}(a)l + \underline{i}_2 \underline{R}_E = -j\omega\mu(\underline{i}_1 + \underline{i}_2)/h \, 2al \, ; \quad \underline{E}(-c) = -\underline{E}(c) \, .$$

Now with the relations

$$\underline{E}(-a) = -\underline{E}(a) \, ; \qquad \underline{E}(c) = \frac{1}{\kappa} \frac{d\underline{H}}{dx} \bigg|_{\substack{c<x<d \\ x \to c}} = \frac{\underline{i}_1 \alpha}{h\kappa} \coth(\alpha(c-d))$$

$$\underline{E}(a) = \frac{1}{\kappa} \frac{d\underline{H}}{dx} \bigg|_{\substack{a<x<b \\ x \to a}} = \frac{\alpha}{\kappa h} \frac{(\underline{i}_1 + \underline{i}_2) \cosh(\alpha(a-b)) - \underline{i}_1}{\sinh(\alpha(a-b))}$$

$$\int_a^b \underline{H}(x)dx = (\underline{i}_1 + \underline{i}_2)/h \frac{1 - \cosh(\alpha(a-b))}{\alpha \sinh(\alpha(a-b))} + \underline{i}_1/h \frac{\cosh(\alpha(b-a)) - 1}{\alpha \sinh(\alpha(b-a))}$$

$$= \frac{2\underline{i}_1 + \underline{i}_2}{h\alpha} \frac{1 - \cosh(\alpha(a-b))}{\sinh(\alpha(a-b))}$$

we can identify two equations for the determination of the complex amplitudes $\underline{i}_1$ and $\underline{i}_2$.

$$-2\underline{i}_1 \frac{\alpha l}{h\kappa} \coth(\alpha(c-d)) + j\omega\mu \left[ \frac{2\underline{i}_1}{h}(c-b)\,l + \frac{2(\underline{i}_1 + \underline{i}_2)}{h}al + \right.$$

$$\left. + (2\underline{i}_1 + \underline{i}_2) \frac{2l}{h\alpha} \frac{1 - \cosh(\alpha(a-b))}{\sinh(\alpha(a-b))} \right] + \underline{i}_1(\underline{R}_G + \underline{R}_S) = \underline{U}_0$$

$$2\frac{\alpha l}{\kappa h} \frac{\underline{i}_1 - (\underline{i}_1 + \underline{i}_2) \cosh(\alpha(a-b))}{\sinh(\alpha(a-b))} + j\omega\mu \frac{(\underline{i}_1 + \underline{i}_2)}{h} 2al + \underline{i}_2 \underline{R}_E = 0$$

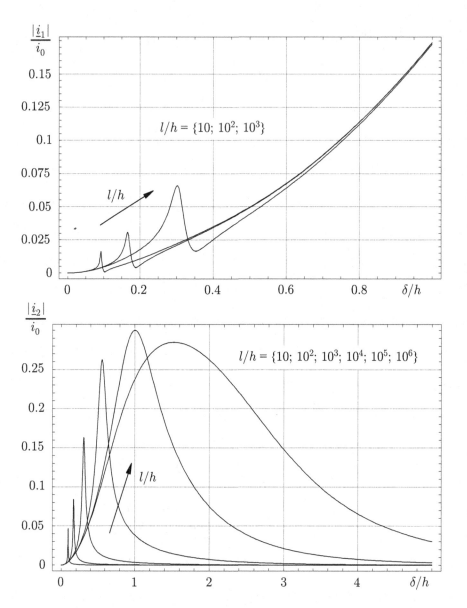

Fig. 5.13–1: Magnitude of the normalized primary current $\underline{i}_1/i_0$ (top) and secondary current $\underline{i}_2/i_0$ (bottom) in dependence on the normalized skin depth $\delta/h$ with $2a/h = 1$, $\underline{R}_G = \underline{R}_S = 0$, and $\underline{R}_E = 1/(j\omega C)$ with $C = 10^{-5}$ [F] $(i_0 = \underline{i}_1|_{\omega \to 0})$

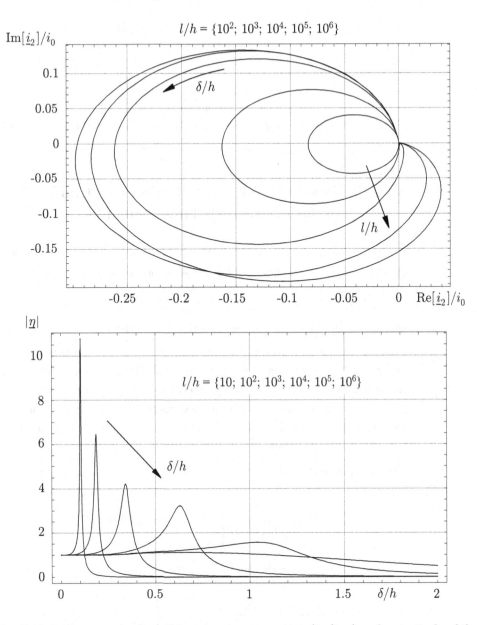

Fig. 5.13–2: Parametric plot of the secondary current $\underline{i}_2/i_0$ (top) and magnitude of the ratio $|\underline{\eta}| = |\underline{i}_2/\underline{i}_1|$ (bottom) in dependence on the normalized skin depth $\delta/h$

Parameters see Fig. 5.13–1

In the limit $\omega = 0$ the following equations are valid

$$U_0 = i_1(R_G + R_S) + i_1 \frac{2l}{\kappa h(d - c)}$$

$$i_2(R_E + \frac{2l}{\kappa h(b - a)}) = 0 ; \qquad i_2 = 0 ,$$

where the complex amplitudes have been replaced by the corresponding real amplitudes.

## 5.14   Induced Current Distribution in a Conducting Slab with Arbitrary Time-Dependency

Consider a slab in $-a \leq x \leq a$ with conductivity $\kappa$ and permeability $\mu$. The infinite slab is exposed to the field of two time-dependent current sheets

$$\vec{K} = \mp \vec{e}_z K(t)$$

on planes $x = \pm b$ with $b > a$ in $z$-direction. The space $|x| > a$ has the permeability $\mu_0$

Find the magnetic field and the current density inside the conducting slab in case of an arbitrary time-dependency of the exciting current sheets. The dissipated power in the conductor for $t \geq 0$ has to be determined, when direct currents are turned off at the time $t = 0$. Compare the dissipated energy with the energy of the magnetic field stored in the conducting material before the turn-off. Finally analyze the case of a harmonic time-dependency.

The exciting field $\vec{H}_E(x,t)$ of the current sheets without presence of the slab is homogeneous in $|x| < b$.

$$\vec{H}_E(x,t) = \begin{cases} \vec{e}_y H_E(t) = \vec{e}_y K(t) & ; \quad |x| < b \\ 0 & ; \quad |x| > b \end{cases}$$

In presence of the slab the field is described by the vector potential

$$\vec{A} = \vec{e}_z A(x,t) ; \qquad \Delta A = \frac{\partial^2 A}{\partial x^2} = \begin{cases} \kappa\mu\partial A/\partial t & ; \quad |x| \leq a \\ 0 & ; \quad |x| > a; \ x \neq \pm b \end{cases} .$$

The solution for an arbitrary time-dependence of the current sheet $K(t)$ can be obtained from the solution for the turn-off of a constant current sheet $K_0$.

$$K(t) = \begin{cases} K_0 & ; \quad t < 0 \\ 0 & ; \quad t \geq 0 \end{cases}$$

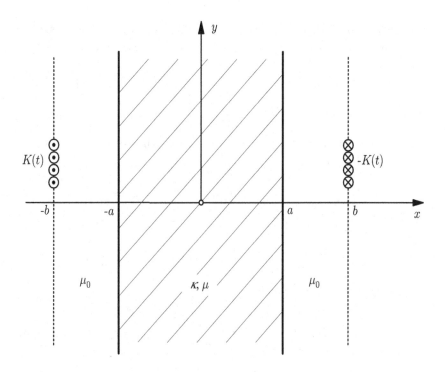

In this case the differential equation is solved for $t \geq 0$ by a separation of variables $A(x,t) = X(x)T(t)$. This results in two ordinary differential equations

$$\frac{1}{X} \frac{d^2 X}{dx^2} = \frac{\kappa \mu}{T} \frac{dT}{dt} = -p^2 ; \qquad |x| \leq a$$

with the solutions $\cos(px), \sin(px)$, and $\exp(-p^2/(\kappa\mu)t)$.

The magnetic field

$$\vec{H} = \frac{1}{\mu} \text{rot}\, \vec{A} = -\vec{e}_y \frac{1}{\mu} \frac{\partial A}{\partial x} = \vec{e}_y\, H(x,t) ; \qquad |x| \leq a$$

satisfies the relation $H(-x) = H(x)$ and thus the vector potential takes the form

$$A(x,t) = \sum_p c_p \sin(px) \exp(-p^2/(\kappa\mu)t) ; \qquad |x| \leq a .$$

With it the magnetic field is

$$H(x,t) = -\frac{1}{\mu} \sum_p p\, c_p \cos(px) \exp(-p^2/(\kappa\mu)t) ; \qquad |x| \leq a .$$

At the time $t \to 0$ it has to satisfy the condition

$$H(x,t \to 0) = -\frac{1}{\mu} \sum_p p\, c_p \cos(px) = \begin{cases} K_0 & |x| < a \\ 0 & x = \pm a \end{cases} ,$$

when the field in $a < |x| < b$ has already vanished, but inside the slab remains unchanged. With $\cos(pa) = 0$ and $p_n a = (2n+1)\pi/2$ with $n = 0, 1, 2, \ldots$ it follows

$$\sum_{n=0}^{\infty} p_n c_n \cos(p_n x) = \begin{cases} -\mu K_0 & ; \quad |x| < a \\ 0 & ; \quad x = \pm a \end{cases}.$$

Now because of the orthogonality of the trigonometric functions the constants $c_n$ are given by

$$p_n c_n a = -\mu K_0 \left. \frac{\sin(p_n x)}{p_n} \right|_{-a}^{a} = \frac{-2\mu K_0}{p_n} \sin(p_n a) = -\frac{2\mu K_0}{p_n}(-1)^n$$

$$\Rightarrow \quad c_n = (-1)^{n+1} \frac{2\mu K_0}{p_n^2 a}.$$

Hence the magnetic field is

$$H(x,t) = 2K_0 \sum_{n=0}^{\infty} \frac{(-1)^n}{p_n a} \cos(p_n x) \exp(-t/\tau_n) ; \qquad \tau_n = \frac{\kappa \mu}{p_n^2}$$

and the current density becomes

$$\text{rot } \vec{H} = \vec{J} = \text{rot } [\vec{e}_y H(x,t)] = \vec{e}_z \frac{\partial H}{\partial x} = \vec{e}_z J(x,t)$$

$$\Rightarrow \quad J(x,t) = \frac{2K_0}{a} \sum_{n=0}^{\infty} (-1)^{n+1} \sin(p_n x) \exp(-t/\tau_n).$$

The dissipated power per unit lengths $l_y$ and $l_z$ in $y$- and $z$-direction after turning of the currents is given by

$$\frac{P_v(t)}{l_y l_z} = \int_{-a}^{a} J(x,t) E(x,t) dx = \frac{1}{\kappa} \int_{-a}^{a} J^2(x,t) dx$$

and the calculation of the dissipated energy results in

$$\frac{W_v}{l_y l_z} = \frac{4K_0^2}{\kappa a^2} \sum_{n=0}^{\infty} \int_{-a}^{a} \sin^2(p_n x) dx \int_{t=0}^{\infty} \exp(-2t/\tau_n) dt$$

$$= \frac{4K_0^2}{\kappa a^2} \sum_{n=0}^{\infty} a \left. \frac{\exp(-2t/\tau_n)}{-2/\tau_n} \right|_0^{\infty} = \frac{2K_0^2}{a} \sum_{n=0}^{\infty} \frac{\mu}{p_n^2} = \frac{2\mu K_0^2}{a} \sum_{n=0}^{\infty} p_n^{-2}.$$

Before the turn-off the energy of the magnetic field stored in $|x| < a$ per unit lengths is

$$\frac{W_m}{l_y l_z} = \frac{1}{2} \int_{-a}^{a} H B dx = \frac{\mu}{2} 4 K_0^2 \sum_{n=0}^{\infty} \frac{1}{(p_n a)^2} \int_{-a}^{a} \cos^2(p_n x) dx$$

$$= \frac{2\mu K_0^2}{a} \sum_{n=0}^{\infty} p_n^{-2} = \frac{W_v}{l_y l_z}$$

and thus is identical to the dissipated energy $W_v$.

In case of an arbitrary time-dependence of the exciting current sheets $K(t)$ it is possible to calculate the resulting field by an integration of infinitesimal field variations. For this purpose the exponential functions $\exp(-t/\tau_n)$ have to be replaced by the functions

$$h_n(t) = \frac{K(0)}{K_0} [1 - \exp(-t/\tau_n)] + \frac{1}{K_0} \int_0^t \frac{dK(t-\tau)}{dt} (1 - \exp(-\tau/\tau_n)) d\tau$$

in the previous expressions.

As an example we consider the excitation of a current sheet with sinusoidal time-dependency.

$$K(t) = \begin{cases} K_0 \cos(\omega t) & ; \quad t \geq 0 \\ 0 & ; \quad t < 0 \end{cases}$$

Then the function $h_n(t)$ takes the form

$$h_n(t) = 1 - \exp(-t/\tau_n) + \int_0^t \frac{d}{dt} \text{Re} \left\{ \exp(j\omega(t-\tau)) \right\} (1 - \exp(-\tau/\tau_n)) d\tau$$

$$= 1 - \exp(-t/\tau_n) + \text{Re} \left\{ \exp(j\omega t) - 1 + \frac{j\omega\tau_n}{1+j\omega\tau_n} \left( \exp(-t/\tau_n) - \exp(j\omega t) \right) \right\}$$

$$= \text{Re} \left\{ \frac{p_n^2}{p_n^2 + \alpha^2} \left( \exp(j\omega t) - \exp(-t/\tau_n) \right) \right\} ; \qquad \alpha^2 = j\omega\kappa\mu .$$

For times $t \gg \tau_n$ the steady state solution remains and the magnetic field can be described by the complex amplitude

$$\underline{H}(x) = 2K_0 \sum_{n=0}^{\infty} \frac{(-1)^n (p_n a)}{(p_n a)^2 + (\alpha a)^2} \cos(p_n x) ; \qquad |x| < a .$$

This result is also obtained from the differential equation for the complex amplitude.

$$\underline{\vec{H}}(x) = \vec{e}_y \underline{H}(x) ; \qquad \frac{d^2 \underline{H}}{dx^2} = \alpha^2 \underline{H} ; \qquad \underline{H}(x) = \underline{C} \frac{\cosh(\alpha x)}{\cosh(\alpha a)}$$

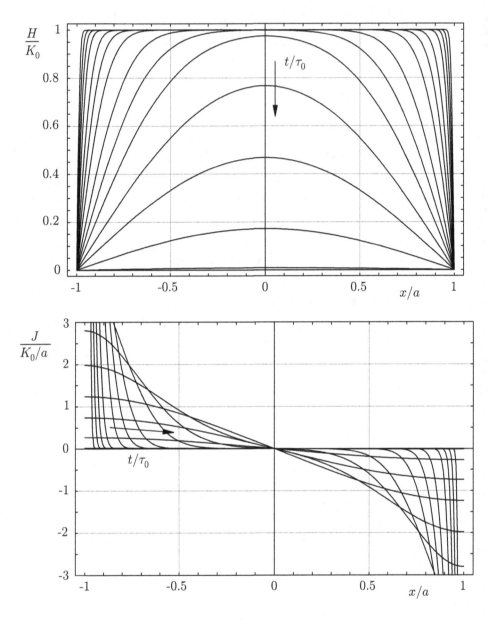

Fig. 5.14–1: Magnetic field $H/K_0$ and current density $J/(K_0/a)$ at different times $t/\tau_0$ after turning off the current sheets

$t/\tau_0 = 2\cdot10^{-4}; 5\cdot10^{-4}; 10^{-3}; 2\cdot10^{-3}; 5\cdot10^{-3}; 10^{-2}; 2\cdot10^{-2}; 5\cdot10^{-2}; 10^{-1}; 2\cdot10^{-1}; 5\cdot10^{-1}; 1; 2; 5; 10$

With $\underline{H}(x = \pm a) = K_0$ it follows

$$\underline{H}(x) = K_0 \frac{\cosh(\alpha x)}{\cosh(\alpha a)} .$$

Now the series expansion of the hyperbolic function

$$\frac{\cosh(\alpha x)}{\cosh(\alpha a)} = \sum_{n=0}^{\infty} b_n \cos(p_n x)$$

leads to

$$b_n a = \int_{-a}^{a} \frac{\cosh(\alpha x)}{\cosh(\alpha a)} \cos(p_n x) dx \;; \qquad b_n = (-1)^n \frac{2 p_n a}{(\alpha a)^2 + (p_n a)^2}$$

and thus

$$\underline{H}(x) = 2 K_0 \sum_{n=0}^{\infty} \frac{(-1)^n (p_n a)}{(\alpha a)^2 + (p_n a)^2} \cos(p_n x)$$

agrees with the previous result.

# 6. Electromagnetic Waves

## 6.1 Transient Waves on Ideal Transmission Lines

Two ideal transmission lines with inductance $L_1'$ and capacitance $C_1'$ in $z > 0$ and $L_2'$ and $C_2'$ in $z < 0$, per unit length respectively, that are of infinite length in one direction on the $z$-axis, are interconnected by a switch $S$ in $z = 0$ at the time $t = 0$. For times $t < 0$ the voltage on the transmission line in $z > 0$ is $U_0$ and on the transmission line in $z < 0$ the voltage is zero.

Calculate the amplitudes of the partial waves (the voltages $u_{1,2}(z,t)$ and the currents $i_{1,2}(z,t)$) in $z > 0$ and $z < 0$ after the switch has been turned on. Furthermore check the invariance of the total energy for times $t > 0$.

After the interconnection at the time $t = 0$ the partial waves $u_1(z - v_1 t)$; $i_1(z - v_1 t)$ and $u_2(z + v_2 t)$; $i_2(z + v_2 t)$ propagate in positive and negative direction with phase velocities $v_{1,2} = (L_{1,2}' C_{1,2}')^{-1/2}$ starting in $z = 0$. For the amplitudes of the partial waves the relations

$$u_1/i_1 = Z_1 = \sqrt{L_1'/C_1'} ; \qquad -u_2/i_2 = Z_2 = \sqrt{L_2'/C_2'}$$

hold. The boundary conditions in $z = 0$ for $t > 0$ require

$$u_2 = U_0 + u_1 ; \qquad i_2 = i_1 = u_1/Z_1 = -u_2/Z_2 ; \qquad z = 0$$

and thus lead to

$$U_0 + u_1 = -u_1 Z_2/Z_1 ; \qquad u_1 = -\frac{U_0}{1 + Z_2/Z_1} ; \qquad u_2 = \frac{U_0}{1 + Z_1/Z_2} .$$

Thus the time responses of the voltages and currents are

$$u(z,t) = U_0 \begin{cases} \dfrac{Z_2}{Z_1 + Z_2} \, \sigma \, (z + v_2 t) & ; \quad z < 0 \\[3mm] 1 - \dfrac{Z_1}{Z_1 + Z_2} \, [1 - \sigma \, (z - v_1 t)] & ; \quad z > 0 \end{cases}$$

$$i(z,t) = -\frac{U_0}{Z_1 + Z_2} \begin{cases} \sigma \, (z + v_2 t) & ; \quad z < 0 \\[3mm] [1 - \sigma \, (z - v_1 t)] & ; \quad z > 0 \end{cases}$$

with

$$\sigma(x) = \begin{cases} 0 & ; \quad x \leq 0 \\[2mm] 1 & ; \quad x > 0 . \end{cases}$$

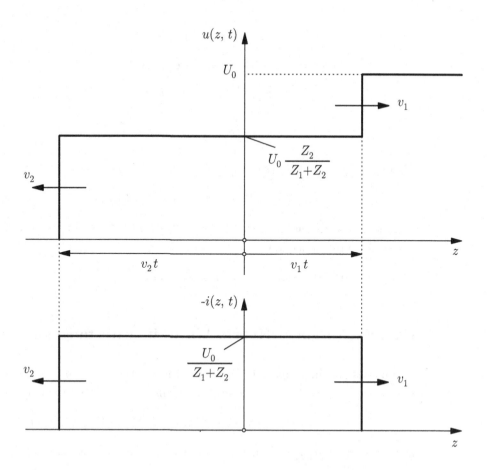

Fig. 6.1–1: Voltage and current in dependence on $z$ and $t$

The energy $W_e(t) + W_m(t)$ of the electromagnetic field stored in the section $-v_2 t < z < v_1 t$ must be equal to the electrostatic energy, that has been stored in $0 < z < v_1 t$ before the interconnection. The energy $W_e(t)$ of the electric field for $t > 0$ is

$$
\begin{aligned}
W_e(t) &= 1/2\, C_1' (U_0 + u_1)^2\, v_1 t + 1/2\, C_2'\, u_2^2\, v_2 t \\
&= 1/2\, C_1'\, U_0^2 \left(1 - \frac{Z_1}{Z_1 + Z_2}\right)^2 (L_1' C_1')^{-1/2}\, t + \\
&\qquad\qquad + 1/2 C_2'\, U_0^2 \left(\frac{Z_2}{Z_1 + Z_2}\right)^2 (L_2' C_2')^{-1/2}\, t \\
&= 1/2\, U_0^2\, t\, \frac{Z_2/Z_1}{Z_1 + Z_2}
\end{aligned}
$$

and the energy $W_m(t)$ of the magnetic field is

$$
\begin{aligned}
W_m(t) &= 1/2\, L_1' i_1^2\, v_1 t + 1/2\, L_2' i_2^2\, v_2 t \\
&= 1/2\, \frac{U_0^2}{(Z_1 + Z_2)^2}\left(L_1'\,(L_1'\,C_1')^{-1/2} + L_2'\,(L_2'\,C_2')^{-1/2}\right) t \\
&= 1/2\, \frac{U_0^2}{Z_1 + Z_2}\, t\,.
\end{aligned}
$$

Therewith the energy of the propagating waves becomes

$$
W_e(t) + W_m(t) = 1/2\, U_0^2/Z_1 t
$$

and this is also the energy, that has been stored in the electrostatic field in $0 < z < v_1$ before the interconnection.

$$
W_e(t < 0)|_{0 < z < v_1 t} = 1/2\, C_1'\, U_0^2\,(L_1'\,C_1')^{-1/2}\, t = 1/2\, U_0^2/Z_1 t
$$

## 6.2    Excitation of Hybrid Waves in a Rectangular Waveguide

The modes of a homogeneous waveguide with rectangular cross section of dimensions $a$ and $b > a$ and with perfect conducting walls are excited by the current

$$
i(y,t) = I_0\,\cos(\pi y/b)\cos(\omega t) = \mathrm{Re}\{I_0\,\cos(\pi y/b)\,\exp(j\omega t)\}
$$

at position $(z = 0\,;\, x = a/2)$. The permittivity $\varepsilon$ and permeability $\mu$ are constant.

Calculate the amplitudes of the excited modes.

The electromagnetic field is described by the vector potentials $\underline{\vec{A}}_{TM} = \vec{e}_z\,\underline{A}_{TM}$ and $\underline{\vec{F}}_{TE} = \vec{e}_z\,\underline{F}_{TE}$.

$$
\underline{\vec{H}} = \underline{\vec{H}}_{TE} + \underline{\vec{H}}_{TM} = \frac{1}{j\omega\mu\varepsilon}\,\mathrm{rot}\,\mathrm{rot}\,\underline{\vec{F}}_{TE} + \frac{1}{\mu}\,\mathrm{rot}\,\underline{\vec{A}}_{TM}
$$

$$
\underline{\vec{E}} = \underline{\vec{E}}_{TE} + \underline{\vec{E}}_{TM} = -1/\varepsilon\,\mathrm{rot}\,\underline{\vec{F}}_{TE} + \frac{1}{j\omega\mu\varepsilon}\,\mathrm{rot}\,\mathrm{rot}\,\underline{\vec{A}}_{TM}
$$

$$
\underline{\vec{B}}_{TM} = \mathrm{rot}\,(\underline{\vec{A}}_{TM}) = \mathrm{grad}\,\underline{A}_{TM} \times \vec{e}_z = -\vec{e}_y\,\frac{\partial \underline{A}_{TM}}{\partial x} + \vec{e}_x\,\frac{\partial \underline{A}_{TM}}{\partial y}
$$

$$
\begin{aligned}
\underline{\vec{E}}_{TM} &= \frac{1}{j\omega\mu\varepsilon}\,\mathrm{rot}\,\mathrm{rot}\,\underline{\vec{A}}_{TM} = \\
&= \frac{1}{j\omega\mu\varepsilon}\left[\vec{e}_x\,\frac{\partial^2 \underline{A}_{TM}}{\partial x \partial z} + \vec{e}_y\,\frac{\partial^2 \underline{A}_{TM}}{\partial y \partial z} + \vec{e}_z\left(\beta^2 - (\beta_{zmn}^{TM})^2\right)\underline{A}_{TM}\right]
\end{aligned}
$$

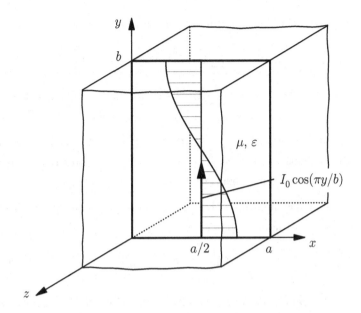

$$\vec{\underline{E}}_{TE} = -\frac{1}{\varepsilon}\,\mathrm{rot}\,\vec{\underline{F}}_{TE} = \frac{1}{\varepsilon}\left(\vec{e}_y\,\frac{\partial \underline{F}_{TE}}{\partial x} - \vec{e}_x\,\frac{\partial \underline{F}_{TE}}{\partial y}\right)$$

$$\vec{\underline{H}}_{TE} = \frac{1}{j\omega\mu\varepsilon}\mathrm{rot\,rot}\,\vec{\underline{F}}_{TE} =$$

$$= \frac{1}{j\omega\mu\varepsilon}\left[\vec{e}_x\,\frac{\partial^2 \underline{F}_{TE}}{\partial x\partial z} + \vec{e}_y\,\frac{\partial^2 \underline{F}_{TE}}{\partial y\partial z} + \vec{e}_z\left(\beta^2 - \left(\beta_{z_{mn}}^{TE}\right)^2\right)\underline{F}_{TE}\right]$$

$$\beta^2 = \omega^2\mu\varepsilon\ ;\qquad \beta_{z_{mn}}^2 = \beta^2 - \beta_{mn}^2$$

The potentials satisfy the differential equations

$$\Delta\,\underline{A}_{TM} + \beta^2\,\underline{A}_{TM} = 0\ ;\qquad \Delta\,\underline{F}_{TE} + \beta^2\,\underline{F}_{TE} = 0\,.$$

At the perfect conducting boundary with contour $C$ and surface normal $\vec{n}$ the tangential component of the electric field $\vec{E}$ and the normal component of the magnetic field $\vec{B} = \mu\,\vec{H}$ vanish. This requirement leads to the following conditions for the potentials.

$$\underline{A}_{TM}|_C = 0\ ;\qquad \left.\frac{\partial \underline{F}_{TE}}{\partial n}\right|_C = 0$$

The excited waves propagate in positive and negative $z$-direction starting from the point of excitement in $z = 0$. With respect to the boundary conditions the solution of the differential equation for the potentials is

$$\underline{A}_{TM} = \sum_{m,n} \underline{C}_{mn}^{(1,2)} \sin(m\pi\, x/a)\, \sin(n\pi\, y/b)\, \exp(\mp j\beta_{z_{mn}}^{TM} z)$$

$$\underline{F}_{TE} = \sum_{m,n} \underline{A}_{mn}^{(1,2)} \cos(m\pi\, x/a)\, \cos(n\pi\, y/b)\, \exp(\mp j\beta_{z_{mn}}^{TE} z)\,.$$

Here the constants $\underline{C}_{mn}^{(1)}$ and $\underline{A}_{mn}^{(1)}$ ($\underline{C}_{mn}^{(2)}$ and $\underline{A}_{mn}^{(2)}$) belong to the waves that propagate in positive (negative) $z$-direction according to the negative (positive) sign in the argument of the exponential function.

As the eigenvalues of the rectangular waveguide are degenerated applies for the propagation constants

$$\beta_{z_{mn}}^{TE} = \beta_{z_{mn}}^{TM} = \beta_{z_{mn}} = \left[\beta^2 - \beta_{mn}^2\right]^{1/2}\,;\qquad \beta_{mn}^2 = (m\,\pi/a)^2 + (n\,\pi/b)^2\,.$$

For the evaluation of the boundary conditions in $z = 0$ it is convenient to consider a current-sheet with a complex amplitude $\underline{K}(x,y)$.

The $z$-component of the magnetic flux density

$$\vec{e}_z \vec{B} = (\beta^2 - \beta_{z_{mn}}^2)\, \underline{F}_{TE}/(j\omega\varepsilon)$$

is continuous in $z = 0$. Hence it is

$$\underline{A}_{mn}^{(1)} = \underline{A}_{mn}^{(2)} = \underline{A}_{mn}\,.$$

For the electric flux density $\vec{D} = \varepsilon\,\vec{E}$ holds the condition

$$\vec{e}_z \left( \vec{D}\Big|_{\substack{z>0 \\ z\to 0}} - \vec{D}\Big|_{\substack{z<0 \\ z\to 0}} \right) = \sigma\,.$$

Because of the symmetry with respect to $z = 0$ this boundary condition leads to

$$\underline{C}_{mn}^{(1)} = -\underline{C}_{mn}^{(2)} = \underline{C}_{mn}.$$

The continuity of the tangential component of the electric field $\vec{E}$ in $z = 0$ gives the same result. Finally the last boundary condition in $z = 0$ is

$$\vec{e}_z \times \left( \vec{\underline{H}}\Big|_{\substack{z>0 \\ z\to 0}} - \vec{\underline{H}}\Big|_{\substack{z<0 \\ z\to 0}} \right) = \vec{\underline{K}}(x,y) = \vec{e}_y\, \underline{K}(x,y)$$

$$\underline{H}_y\Big|_{\substack{z>0 \\ z\to 0}} - \underline{H}_y\Big|_{\substack{z<0 \\ z\to 0}} = 0\,;\qquad \underline{H}_x\Big|_{\substack{z>0 \\ z\to 0}} - \underline{H}_x\Big|_{\substack{z<0 \\ z\to 0}} = \underline{K}(x,y)$$

$$\underline{H}_y = -\frac{1}{\mu}\frac{\partial \underline{A}_{TM}}{\partial x} + \frac{1}{j\omega\mu\varepsilon}\frac{\partial^2 \underline{F}_{TE}}{\partial y \partial z}$$

$$\underline{H}_x = \frac{1}{\mu}\frac{\partial \underline{A}_{TM}}{\partial y} + \frac{1}{j\omega\mu\varepsilon}\frac{\partial^2 \underline{F}_{TE}}{\partial x \partial z}.$$

At first the $y$-component of the magnetic field yields

$$\sum_{m,n} \cos(m\pi x/a)\sin(n\pi y/b)\left[-\frac{2}{\mu}\underline{C}_{mn}\frac{m\pi}{a} + \frac{2n\pi j\beta_{zmn}}{j\omega\mu\varepsilon b}\underline{A}_{mn}\right] = 0$$

$$\Rightarrow \quad \underline{C}_{mn} = \frac{\beta_{zmn}}{\omega\varepsilon}\frac{n}{m}\frac{a}{b}\underline{A}_{mn}$$

and with it the evaluation of the $x$-component results in

$$\sum_{m,n} \sin(m\pi x/a)\cos(n\pi y/b)\left[\frac{1}{\mu}\frac{n\pi}{b}2\underline{C}_{mn} + \frac{2m\pi j\beta_{zmn}}{j\omega\mu\varepsilon a}\underline{A}_{mn}\right] = \underline{K}(x,y)$$

$$\frac{2}{\mu}\sum_{m,n}\frac{\beta_{zmn}}{\omega\varepsilon}\frac{m\pi}{a}\underline{A}_{mn}\left[1 + (n/m)^2(a/b)^2\right]\sin(m\pi x/a)\cos(n\pi y/b) =$$

$$= \underline{K}(x,y) = I_0\cos(\pi y/b)\,\delta(x - a/2)\,.$$

Thus because of the given exciting current only solutions with $n = 1$ exist and with the orthogonality of the trigonometric functions we obtain after multiplication with $\sin(k\pi x/a)$ and integration from $x = 0$ to $x = a$

$$\frac{2}{\mu}\frac{\beta_{zm1}}{\omega\varepsilon}\frac{m\pi}{a}\underline{A}_{m1}\left[1 + (a/b)^2/m^2\right]a/2 = I_0\sin(m\pi/2) = I_0\begin{cases} 0 & ;\quad m = 2k \\ (-1)^k & ;\quad m = 2k+1 \end{cases}$$

with $k = 0, 1, 2, \ldots$.

The final result is

$$\underline{A}_{2k+1,1} = \frac{\mu I_0}{\pi}\frac{(-1)^k}{1 + (a/b)^2/m^2}\frac{\omega\varepsilon}{\beta_{z2k+1,1}}\frac{1}{2k+1}\;;\qquad \underline{A}_{2k,1} = 0$$

$$\underline{C}_{2k+1,1} = \frac{\mu I_0}{\pi}\frac{(-1)^k}{1 + (a/b)^2/m^2}\frac{a/b}{(2k+1)^2}\;;\qquad \underline{C}_{2k,1} = 0\;;\qquad k = 0, 1, 2, \ldots\,.$$

## 6.3   Excitation of Transverse Electric Waves
in a Parallel-Plate Waveguide

The field of a short-circuited parallel-plate waveguide with perfect conducting boundaries in $y = 0$, $y = a$, and $z = 0$ is excited by a current sheet

$$\vec{K}(y,t) = \vec{e}_x K_0 \cos(\pi y/a) \cos(\omega t) = \vec{e}_x \text{Re}\left\{K_0 \cos(\pi y/a) \exp(j\omega t)\right\}$$

in $(z = b\,;\, 0 < y < a)$. Inside the waveguide $(0 < y < a\,;\, z > 0)$ the permittivity is $\varepsilon$
and the permeability is $\mu$.

Calculate the amplitudes of the excited Modes.

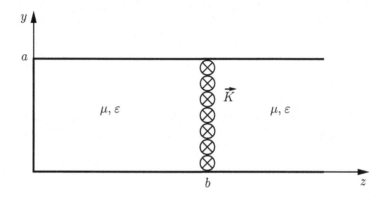

The field inside the waveguide is divided into transverse electric (TE-) and transverse
magnetic (TM-)Waves with respect to the $z$-direction and is thus described by the
vector potentials

$$\vec{\underline{A}}_{TM} = \vec{e}_z \underline{A}_{TM}\,; \qquad \Delta \vec{\underline{A}}_{TM} + \beta^2 \vec{\underline{A}}_{TM} = 0$$

$$\vec{\underline{F}}_{TE} = \vec{e}_z \underline{F}_{TE}\,; \qquad \Delta \vec{\underline{F}}_{TE} + \beta^2 \vec{\underline{F}}_{TE} = 0$$

in $z \neq b$ with $\beta^2 = \omega^2 \mu \varepsilon$. Now the field is derived from

$$\vec{\underline{H}}_{TM} = \frac{1}{\mu}\text{rot}\,(\vec{\underline{A}}_{TM}) = \frac{1}{\mu}\text{grad}\,\underline{A}_{TM} \times \vec{e}_z = \frac{1}{\mu}\left[-\vec{e}_y \frac{\partial \underline{A}_{TM}}{\partial x} + \vec{e}_x \frac{\partial \underline{A}_{TM}}{\partial y}\right]$$

$$\vec{\underline{E}}_{TM} = \frac{1}{j\omega\mu\varepsilon}\text{rot}\,\text{rot}\,\vec{\underline{A}}_{TM} = \frac{1}{j\omega\mu\varepsilon}\left[\vec{e}_x \frac{\partial^2 \underline{A}_{TM}}{\partial x \partial z} + \vec{e}_y \frac{\partial^2 \underline{A}_{TM}}{\partial y \partial z} + \vec{e}_z (\beta^2 - \beta_z^{TM^2})\underline{A}_{TM}\right]$$

$$\underline{\vec{E}}_{TE} = -\frac{1}{\varepsilon}\operatorname{rot}\underline{\vec{F}}_{TE} = \frac{1}{\varepsilon}\left(\vec{e}_y\frac{\partial\underline{F}_{TE}}{\partial x} - \vec{e}_x\frac{\partial\underline{F}_{TE}}{\partial y}\right)$$

$$\underline{\vec{H}}_{TE} = \frac{1}{j\omega\mu\varepsilon}\operatorname{rot}\operatorname{rot}\underline{\vec{F}}_{TE} = \frac{1}{j\omega\mu\varepsilon}\left[\vec{e}_x\frac{\partial^2\underline{F}_{TE}}{\partial x\partial z} + \vec{e}_y\frac{\partial^2\underline{F}_{TE}}{\partial y\partial z} + \vec{e}_z(\beta^2 - \beta_z^{TE^2})\underline{F}_{TE}\right].$$

As the field is independent of the $x$-coordinate, the transverse magnetic field has only a $x$-component and thus is not excited by a current sheet in $x$-direction. Hence the resulting field is purely transverse electric with respect to the $z$-direction.

At the perfect conducting boundary the tangential component of the electric field and the normal component of the magnetic flux density vanish. These conditions are satisfied, when the normal derivative of the potential $\underline{F}_{TE}$ vanishes at the boundaries $y = 0$, $y = a$:

$$\left.\frac{\partial\underline{F}_{TE}}{\partial y}\right|_{y=0} = \left.\frac{\partial\underline{F}_{TE}}{\partial y}\right|_{y=a} = 0,$$

and in $z = 0$ the potential itself vanishes:

$$\underline{F}_{TE}\big|_{z=0} = 0.$$

If only the first condition is met, the solution of the differential equation for the vector potential $\underline{F}_{TE}$ reads

$$\underline{F}_{TE} = \sum_{n=1}^{\infty}\underline{C}_n\cos(n\pi y/a)\exp(\mp j\beta_{zn}^{TE}z);\qquad \beta_{zn}^{TE} = \sqrt{\beta^2 - (n\pi/a)^2} = \beta_{zn}.$$

This solution describes waves, that propagate in positive (negative) $z$-direction, according to the negative (positive) sign in the argument of the exponential function. In $0 < z < b$ we need waves, that propagate in both directions, so that the sum of all waves satisfies the boundary condition in $z = 0$. The resulting vector potential is

$$\underline{F}_{TE} = \sum_{n=1}^{\infty}\left\{\begin{matrix}\underline{C}_n^{(1)} \\ \underline{C}_n^{(2)}\end{matrix}\right\}\cos(n\pi y/a)\left\{\begin{matrix}\sin(\beta_{zn}z)/\sin(\beta_{zn}b) & ; & 0 < z < b \\ \exp(-j\beta_{zn}(z-b)) & ; & z > b.\end{matrix}\right.$$

The continuity of the $z$-component of the magnetic flux density in $z = b$ requires the continuity of the potential $\underline{F}_{TE}$, and hence $\underline{C}_n^{(1)} = \underline{C}_n^{(2)} = \underline{C}_n$.

This constant follows from the boundary condition for the tangential component of the magnetic field at the position of the current sheet.

$$\underline{H}_y\Big|_{\substack{z<b\\z\to b}} - \underline{H}_y\Big|_{\substack{z>b\\z\to b}} = K_0\cos(\pi y/a) = \frac{1}{j\omega\mu\varepsilon}\left[\frac{\partial^2\underline{F}_{TE}}{\partial y\partial z}\Big|_{\substack{z<b\\z\to b}} - \frac{\partial^2\underline{F}_{TE}}{\partial y\partial z}\Big|_{\substack{z>b\\z\to b}}\right] =$$

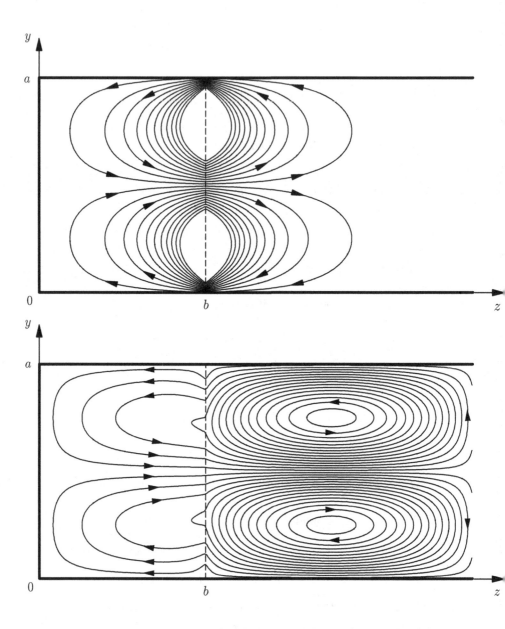

Fig. 6.3–1: Magnetic lines of force below the cut-off ($\omega = 10^{11}\,[\mathrm{s}^{-1}]$, top) and for an
          excitation of one propagating mode
          ($\omega = 2 \cdot 10^{11}\,[\mathrm{s}^{-1}]$, bottom)
          Parameters: $a = 0.01\,[\mathrm{m}]$, $b/a = 0.8$, $\varepsilon = \varepsilon_0$, and $\mu = \mu_0$

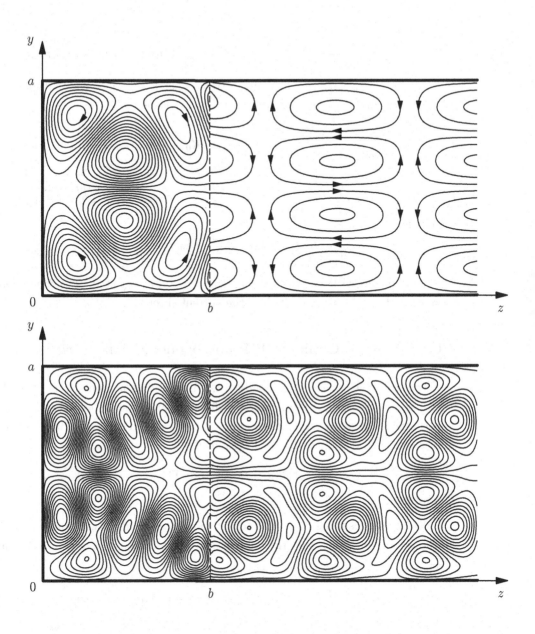

Fig. 6.3–2: Magnetic lines of force when two ($\omega = 4 \cdot 10^{11}$ [s$^{-1}$], top) and three ($\omega = 6 \cdot 10^{11}$ [s$^{-1}$], bottom) propagating modes are excited, Parameters: see Fig. 6.3–1

$$= -\frac{1}{j\omega\mu\varepsilon} \sum_{n=1}^{\infty} \underline{C}_n \frac{n\pi}{a} \beta_{zn} \frac{\exp(j\beta_{zn}b)}{\sin(\beta_{zn}b)} \sin(n\pi y/a) = K_0 \cos(\pi y/a)$$

Now, due to the orthogonality of the trigonometric functions the multiplication with $\sin(k\pi y/a)$ and integration from $y = 0$ to $y = a$ leads to

$$-\frac{1}{j\omega\mu\varepsilon} \underline{C}_k \frac{k\pi}{a} \beta_{zk} \frac{\exp(j\beta_{zk}b)}{\sin(\beta_{zk}b)} \frac{a}{2} = K_0 \cdot \underbrace{\int_0^a \cos(\pi y/a)\sin(k\pi y/a)dy}_{\substack{= 0 \text{ for } k = 2m+1 \text{ and} \\ = a/\pi 2k/(k^2-1) \text{ for } k = 2m}}$$

and thus

$$\underline{C}_n = -jK_0 \frac{4}{\pi^2} \frac{\omega\mu\varepsilon a}{\beta_{zn}} \frac{\sin(\beta_{zn}b)\exp(-j\beta_{zn}b)}{n^2-1}; \quad n = 2m; \quad m = 1, 2, 3, \dots .$$

So only modes with even modal numbers $n = 2m$ are excited. The fundamental TE mode with $n = 1$ is not excited by the current sheet.

## 6.4   Coaxial Cable with Inhomogeneous Dielectric

A coaxial cable with perfect conductors, inner radius $a$, and outer radius $b > a$ is of infinite extension in the negative $z$-direction and is short circuited in $z = c$. The permittivity is $\varepsilon_0$ in $-\infty < z < 0$ and $\varepsilon$ in $0 \le z < c$. The permeability $\mu$ is constant. In $z < 0$ a TEM-wave with the electric field

$$\vec{\underline{E}}_e = \vec{e}_\varrho E_0\, a/\varrho \exp(-j\beta_0 z)\,; \qquad \beta_0 = \omega \sqrt{\mu\varepsilon_0}$$

is excited.

Calculate the resulting field and find the voltage $\underline{u}(z)$ and the current $\underline{i}(z)$ on the cable. Choose a value $c$ such that the voltage $\underline{u}(z)$ at position $z = -\lambda_0 = -2\pi/\beta_0$ vanishes.

The magnetic field of the excited wave is

$$\vec{\underline{H}}_e = 1/Z_0\vec{e}_z \times \vec{\underline{E}}_e = \vec{e}_\varphi E_0/Z_0 a/\varrho \exp(-j\beta_0 z)\,; \qquad Z_0 = \sqrt{\mu/\varepsilon_0}\,.$$

An ansatz for the resulting electric field, that accounts for the reflections at $z = 0$ and $z = c$, is

$$\vec{\underline{E}}(\varrho, z) = \vec{e}_\varrho E_0 \frac{a}{\varrho} \begin{cases} \exp(-j\beta_0 z) + \underline{A}\exp(j\beta_0 z) & ;\quad z \le 0 \\ \underline{B}\exp(-j\beta z) + \underline{C}\exp(j\beta z) & ;\quad 0 \le z \le c \end{cases}$$

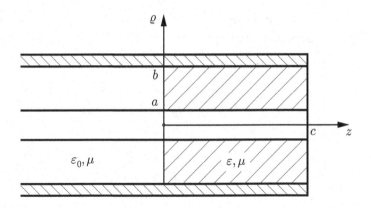

with waves, that propagate in $\pm z$-direction with phase constants $\beta_0$ and $\beta = \omega\sqrt{\mu\varepsilon}$. For each partial wave holds the fundamental relation $\vec{H} = \pm 1/Z\,\vec{e}_z \times \vec{E}$ and thus it is

$$\vec{H}(\varrho, z) = \vec{e}_\varphi \frac{E_0}{Z_0} \frac{a}{\varrho} \begin{cases} \exp(-j\beta_0 z) - \underline{A}\exp(j\beta_0 z) & ; \quad z \le 0 \\ Z_0/Z(\underline{B}\exp(-j\beta z) - \underline{C}\exp(j\beta z)) & ; \quad 0 \le z \le c \end{cases}$$

with $Z = \sqrt{\mu/\varepsilon}$ for $0 < z < c$.

The unknown constants follow from the boundary conditions

$$\vec{e}_z \times \vec{E}\Big|_{z=c} = 0\,; \qquad \underline{E}(\varrho, z)\big|_{\substack{z<0 \\ z\to 0}} = \underline{E}(\varrho, z)\big|_{\substack{z>0 \\ z\to 0}}\,; \qquad \underline{H}(\varrho, z)\big|_{\substack{z<0 \\ z\to 0}} = \underline{H}(\varrho, z)\big|_{\substack{z>0 \\ z\to 0}}\,.$$

This leads to the equations   $1 + \underline{A} = \underline{B} + \underline{C}\,;$

$$1 - \underline{A} = Z_0/Z\,(\underline{B} - \underline{C})\,; \qquad \underline{B}\exp(-j\beta c) + \underline{C}\exp(j\beta c) = 0$$

and finally

$$\underline{A} = \frac{r - \exp(-2j\beta c)}{1 - r\exp(-2j\beta c)}\,; \qquad \underline{B} = \frac{2Z}{Z_0 + Z}\,(1 - r\exp(-2j\beta c))^{-1}$$

$$\underline{C} = \frac{-2Z}{Z_0 + Z}\frac{\exp(-2j\beta c)}{1 - r\exp(-2j\beta c)}\,; \qquad r = \frac{Z - Z_0}{Z + Z_0}\,.$$

In the purely formal limit $Z \to 0$ it is $r = -1$ and

$$\underline{A}\big|_{Z\to 0} = -1\,; \qquad \underline{C}\big|_{Z\to 0} = \underline{B}\big|_{Z\to 0} = 0.$$

This is the solution for the total reflection of incident waves at a perfect conducting wall in $z = 0$. With $\varepsilon = \varepsilon_0$ it is $r = 0$ and

$$\underline{A}\big|_{Z=Z_0} = -\exp(-2j\beta c)\,; \qquad \underline{B}\big|_{Z=Z_0} = 1\,; \qquad \underline{C}\big|_{Z=Z_0} = -\exp(-2j\beta c).$$

Now this describes the total reflection at the plane $z = c$ in a coaxial cable with homogeneous permittivity $\varepsilon_0$.

The voltage and the current on the cable are

$$\underline{u}(z) = \int_a^b \vec{\underline{E}}(\varrho, z)\, \vec{e}_\varrho \, d\varrho = \int_a^b \underline{E}(\varrho, z)\, d\varrho =$$

$$= \underline{u}_0 \, \ln(b/a) \begin{cases} \exp(-j\beta_0 z) + \underline{A} \exp(j\beta_0 z) & ; \quad z \leq 0 \\ \underline{B} \, \exp(-j\beta z) + \underline{C} \exp(j\beta z) & ; \quad 0 \leq z \leq c \end{cases}$$

$$\underline{i}(z) = \int_0^{2\pi} \vec{\underline{H}}(\varrho, z) \vec{e}_\varphi \, \varrho d\varphi$$

$$= 2\pi \underline{i}_0 \begin{cases} \exp(-j\beta_0 z) - \underline{A} \exp(j\beta_0 z) & ; \quad z \leq 0 \\ Z_0/Z(\underline{B} \, \exp(-j\beta z) - \underline{C} \exp(j\beta z)) & ; \quad 0 \leq z \leq c \end{cases}$$

with $\underline{u}_0 = \underline{E}_0 a$ and $\underline{i}_0 = \underline{u}_0/Z_0$.

The constraint $\underline{u}(z = -2\pi/\beta_0) = 0$ is satisfied if $1 + \underline{A} = 0$ and thus

$$\underline{A} = -1 = (r-1)/(1-r)\,.$$

This leads to $\exp(-2j\beta c) = 1$ and results in $c = m\pi/\beta$; $m = 0, 1, 2, \ldots$.

## 6.5 Cylindrical Waveguide Resonator with Inhomogeneous Permittivity

A cylindrical waveguide resonator with perfect conducting boundaries in $\varrho = a$, $z = -c$ and $z = d$ is filled in ($\varrho < a$; $-c < z < 0$) with material of permittivity $\varepsilon_1$ and in ($\varrho < a$; $0 \leq z < d$) with material of permittivity $\varepsilon_2$. The permeability $\mu$ is constant.

Find the resonance frequencies of the modes, that are transverse electric (TE) or transverse magnetic (TM) relative to the $z$-axis.

The field inside the resonator can be derived by means of axial vector potentials

$$\vec{\underline{A}}_{TM} = \vec{e}_z \underline{A}_{TM}\,; \qquad \vec{\underline{F}}_{TE} = \vec{e}_z \underline{F}_{TE}$$

$$\Delta \underline{A}_{TM} + \beta_{1,2}^2 \underline{A}_{TM} = 0\,; \qquad \Delta \underline{F}_{TE} + \beta_{1,2}^2 \underline{F}_{TE} = 0,$$

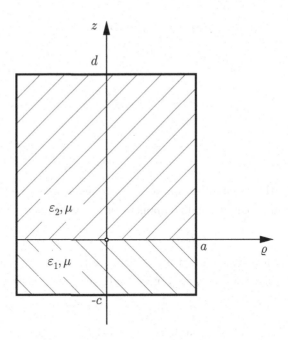

out of which the field is calculated by

$$\vec{B}_{TM} = \operatorname{rot}\underline{\vec{A}}_{TM} = -\vec{e}_z \times \operatorname{grad}\underline{A}_{TM} = \mu\underline{\vec{H}}_{TM}$$

$$\underline{\vec{E}}_{TM} = \frac{1}{j\omega\mu\varepsilon_{1,2}}\left(\operatorname{grad}{}_s(\partial\underline{A}_{TM}/\partial z) + \vec{e}_z(\beta_{1,2}^2 - \beta_{TM}^{(1,2)2})\underline{A}_{TM}\right)$$

$$\underline{\vec{E}}_{TE} = -1/\varepsilon_{1,2}\operatorname{rot}\underline{\vec{F}}_{TE} = 1/\varepsilon_{1,2}\,\vec{e}_z \times \operatorname{grad}\underline{F}_{TE}\,; \quad \operatorname{grad}{}_s = \operatorname{grad} - \vec{e}_z\partial/\partial z$$

$$\underline{\vec{H}}_{TE} = \frac{\vec{B}_{TE}}{\mu} = \frac{1}{j\omega\mu\varepsilon_{1,2}}\left(\operatorname{grad}{}_s(\partial\underline{F}_{TE}/\partial z) + \vec{e}_z(\beta_{1,2}^2 - \beta_{TE}^{(1,2)2})\underline{F}_{TE}\right).$$

At the cylinder wall $\varrho = a$ the tangential component of $\underline{\vec{E}}$ and accordingly the normal component of $\vec{B}$ vanish. Therefore it is

$$\underline{A}_{TM}|_{\varrho=a} = 0\,; \qquad \partial\underline{F}_{TE}/\partial\varrho|_{\varrho=a} = 0$$

and in the domain (1) for $-c < z < 0$ and (2) for $0 < z < d$ the solution may be written in the following form.

$$\underline{A}_{TM}^{(1,2)}(\varrho,\varphi,z) = \underline{C}_{TM}^{(1,2)}\,U_{TM_{nr}}(\varrho)\,\phi_n(\varphi)\,Z_{TM_{nr}}^{(1,2)}(z)$$

$$\underline{F}_{TE}^{(1,2)}(\varrho,\varphi,z) = \underline{C}_{TE}^{(1,2)}\,U_{TE_{nr}}(\varrho)\,\phi_n(\varphi)\,Z_{TE_{nr}}^{(1,2)}(z)$$

$$\partial U_{TE_{nr}}/\partial\varrho|_{\varrho=a} = 0\,; \qquad U_{TM_{nr}}(\varrho=a) = 0$$

$$U_{TM_{nr}}(\varrho) = J_n(x_{nr}\varrho/a) \; ; \qquad U_{TE_{nr}}(\varrho) = J_n(x'_{nr}\varrho/a)$$

$$\phi_n(\varphi) = \left\{ \begin{array}{c} \cos n\varphi \\ \sin n\varphi \end{array} \right\} \; ; \qquad J_n(x_{nr}) = 0 \; ; \qquad J'_n(x'_{nr}) = 0$$

$$Z_{TM_{nr}}^{(1,2)}(z) = \exp(\mp j\beta_{TM_{nr}}^{(1,2)}z) \; ; \qquad Z_{TE_{nr}}^{(1,2)}(z) = \exp(\mp j\beta_{TE_{nr}}^{(1,2)}z)$$

$$\beta_{TM_{nr}}^{(1,2)} = \sqrt{\beta_{1,2}^2 - (x_{nr}/a)^2} \; ; \qquad \beta_{TE_{nr}}^{(1,2)} = \sqrt{\beta_{1,2}^2 - (x'_{nr}/a)^2} \; ; \qquad \begin{array}{l} r = 1,2,\ldots \\ n = 0,1,2,\ldots \end{array}$$

Also at the top and bottom wall $z = d$ and $z = -c$ the tangential component of $\vec{E}$ and accordingly the normal component of $\vec{B}$ vanish and

$$\partial \underline{A}_{TM}/\partial z\big|_{z=-c,d} = 0 \; ; \qquad \underline{E}_{TE}\big|_{z=-c,d} = 0$$

holds. The solution functions, that describe waves propagating in positive and negative $z$-direction, have to be superposed with respect to these boundary conditions. It follows

$$Z_{TM_{nr}}^{(1)}(z) = \cos(\beta_{TM_{nr}}^{(1)}(z+c)) \; ; \qquad Z_{TM_{nr}}^{(2)}(z) = \cos(\beta_{TM_{nr}}^{(2)}(z-d))$$

$$Z_{TE_{nr}}^{(1)}(z) = \sin(\beta_{TE_{nr}}^{(1)}(z+c)) \; ; \qquad Z_{TE_{nr}}^{(2)}(z) = \sin(\beta_{TE_{nr}}^{(2)}(z-d)) \, .$$

In $z = 0$ the tangential components of $\vec{E}$ and $\vec{H}$ are continuous.

$$\underline{C}_{TM}^{(1)} \cos(\beta_{TM_{nr}}^{(1)}c) = \underline{C}_{TM}^{(2)} \cos(\beta_{TM_{nr}}^{(2)}d)$$

$$1/\varepsilon_1 \, \underline{C}_{TM}^{(1)} \, \beta_{TM_{nr}}^{(1)} \sin(\beta_{TM_{nr}}^{(1)}c) = -1/\varepsilon_2 \, \underline{C}_{TM}^{(2)} \, \beta_{TM_{nr}}^{(2)} \sin(\beta_{TM_{nr}}^{(2)}d)$$

$$1/\varepsilon_1 \, \underline{C}_{TE}^{(1)} \sin(\beta_{TE_{nr}}^{(1)}c) = -1/\varepsilon_2 \, \underline{C}_{TE}^{(2)} \sin(\beta_{TE_{nr}}^{(2)}d)$$

$$1/\varepsilon_1 \, \underline{C}_{TE}^{(1)} \, \beta_{TE_{nr}}^{(1)} \cos(\beta_{TE_{nr}}^{(1)}c) = 1/\varepsilon_2 \, \underline{C}_{TE}^{(2)} \, \beta_{TE_{nr}}^{(2)} \cos(\beta_{TE_{nr}}^{(2)}d)$$

Therewith the eigenvalue equations for the TE- and TM-Modes follow from the division of the related equations.

$$\beta_{TM_{nr}}^{(1)} \tan(\beta_{TM_{nr}}^{(1)}c) = -\varepsilon_1/\varepsilon_2 \, \beta_{TM_{nr}}^{(2)} \tan(\beta_{TM_{nr}}^{(2)}d)$$

$$\beta_{TM_{nr}}^{(1)^2} - \beta_{TM_{nr}}^{(2)^2} = \beta_1^2 - \beta_2^2 = \omega_{nr}^2\mu(\varepsilon_1 - \varepsilon_2)$$

$$\tan(\beta_{TE_{nr}}^{(1)}c) = -\frac{\beta_{TE_{nr}}^{(1)}}{\beta_{TE_{nr}}^{(2)}} \tan(\beta_{TE_{nr}}^{(2)}d)$$

$$\beta_{TE_{nr}}^{(1)^2} - \beta_{TE_{nr}}^{(2)^2} = \beta_1^2 - \beta_2^2 = \omega_{nr}^2\mu(\varepsilon_1 - \varepsilon_2)$$

The determination of the resonance frequencies $\omega_{nr}$ requires a numerical solution of the transcendental eigenvalue equations.

## 6.6 Guided Waves in a Parallel-Plate Waveguide with Layered Permittivity

Two perfect conducting walls in $y = 0$ and $y = b$ form a parallel-plate waveguide, that is in $0 < y < a$ filled with material of permittivity $\varepsilon_1$, and in $a < y < b$ with material of permittivity $\varepsilon_2$. The permeability $\mu$ is constant.

Calculate the phase constants of the waves, that are independent of the $x$-coordinate and propagate in $z$-direction.

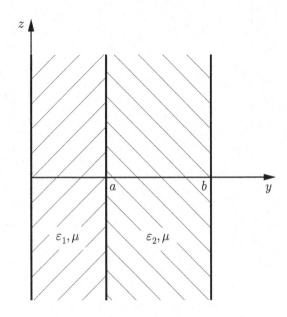

The guided waves are either transverse electric (TE) modes or transverse magnetic (TM) modes. Therefore the field is described by the vector potentials

$$\vec{A}_{TM} = \vec{e}_z \, \underline{A}_{TM}(y,z) \,; \qquad \Delta \underline{A}_{TM} + \beta_{1,2}^2 \underline{A}_{TM} = 0$$

$$\vec{F}_{TE} = \vec{e}_z \, \underline{F}_{TE}(y,z) \,; \qquad \Delta \underline{F}_{TE} + \beta_{1,2}^2 \underline{F}_{TE} = 0 \,; \qquad \beta_{1,2}^2 = \omega^2 \mu \varepsilon_{1,2}$$

with $\quad \underline{\vec{B}}_{TM} = \operatorname{rot} \vec{A}_{TM} \quad$ and $\quad \underline{\vec{E}}_{TE} = -1/\varepsilon \operatorname{rot} \vec{F}_{TE}$.

$$\underline{\vec{B}}_{TM} = \vec{e}_x \frac{\partial \underline{A}_{TM}}{\partial y} \,; \qquad \underline{\vec{E}}_{TM} = \frac{1}{j\omega\mu\varepsilon_{1,2}} \left[ \vec{e}_y \frac{\partial^2 \underline{A}_{TM}}{\partial y \, \partial z} + \vec{e}_z (\beta_{1,2}^2 - \beta_{TM}^2) \underline{A}_{TM} \right]$$

$$\vec{E}_{TE} = -\frac{1}{\varepsilon} \vec{e}_x \frac{\partial \underline{F}_{TE}}{\partial y} \,; \qquad \underline{\vec{H}}_{TE} = \frac{1}{j\omega\mu\varepsilon_{1,2}} \left[ \vec{e}_y \frac{\partial^2 \underline{F}_{TE}}{\partial y \, \partial z} + \vec{e}_z (\beta_{1,2}^2 - \beta_{TE}^2) \underline{F}_{TE} \right]$$

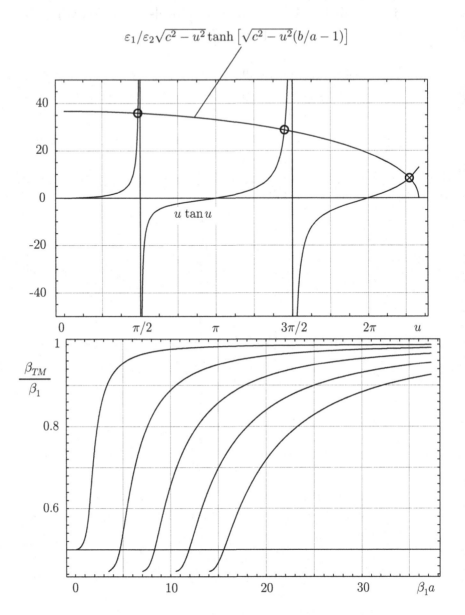

Fig. 6.6–1: Solutions of the eigenvalue equation for TM-modes with
$\omega = 1.1 \cdot 10^{11} \, [\text{s}^{-1}]$ and phase constants in dependence on the normalized
frequency for $\varepsilon_1/\varepsilon_2 = 5$ and $a/b = 0.25$
$c^2 = a^2(\beta_1^2 - \beta_2^2); \; u^2 = a^2(\beta_1^2 - \beta_{TM}^2)$

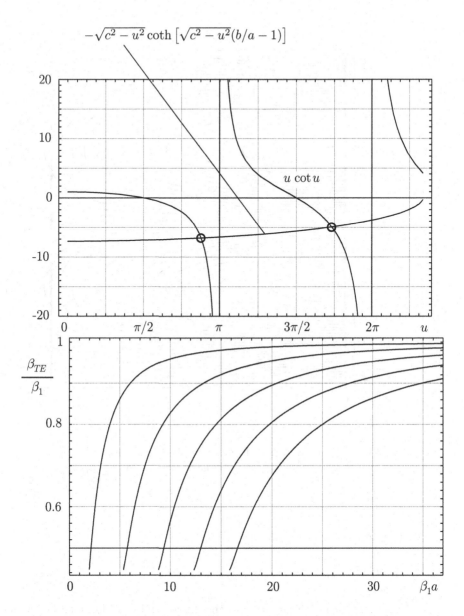

Fig. 6.6–2: Solutions of the eigenvalue equation for TE-modes with
$\omega = 1.1 \cdot 10^{11}$ [s$^{-1}$] and phase constants in dependence on the normalized
frequency for $\varepsilon_1/\varepsilon_2 = 5$ and $a/b = 0.25$
$c^2 = a^2(\beta_1^2 - \beta_2^2)$; $u^2 = a^2(\beta_1^2 - \beta_{TE}^2)$

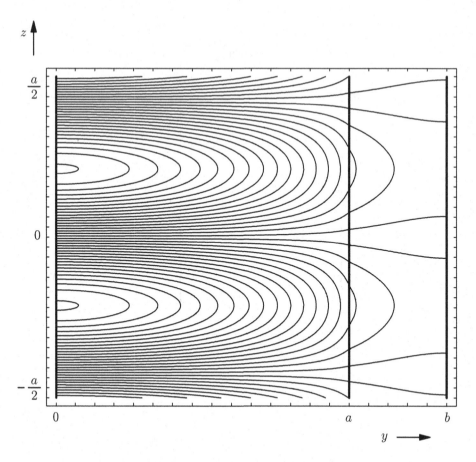

Fig. 6.6–3: Electric lines of force of the fundamental TM-mode
$$\varepsilon_1/\varepsilon_2 = 2; \ \omega = 5 \cdot 10^{10} \ [\text{s}^{-1}]; \ \beta_{TM} = 231.1 \ [\text{m}^{-1}]$$

The field satisfies the boundary conditions

$$\vec{e}_y \times \vec{E}\Big|_{y=0,b} = 0; \quad \vec{e}_y \times \vec{E}\Big|_{\substack{y<a \\ y\to a}} = \vec{e}_y \times \vec{E}\Big|_{\substack{y>a \\ y\to a}}; \quad \vec{e}_y \times \vec{H}\Big|_{\substack{y<a \\ y\to a}} = \vec{e}_y \times \vec{H}\Big|_{\substack{y>a \\ y\to a}} \ .$$

An approach for the solution of the differential equation for the TM-modes is

$$\frac{\partial^2 \underline{A}_{TM}}{\partial y^2} + (\beta_{1,2}^2 - \beta_{TM}^2)\, \underline{A}_{TM} = 0; \qquad \frac{\partial \underline{A}_{TM}}{\partial x} = 0; \qquad \beta_{1,2}^2 = \omega^2 \mu \varepsilon_{1,2}$$

$$\underline{A}_{TM} = \left\{ \begin{array}{l} \underline{C}_1 \sin(\sqrt{\beta_1^2 - \beta_{TM}^2}\,y) \\[2mm] \underline{C}_2 \sin(\sqrt{\beta_2^2 - \beta_{TM}^2}\,(y-b)) \end{array} \right\} \exp(\mp j\beta_{TM}z); \qquad \begin{array}{l} 0 < y < a \\[2mm] a < y < b \end{array} \ .$$

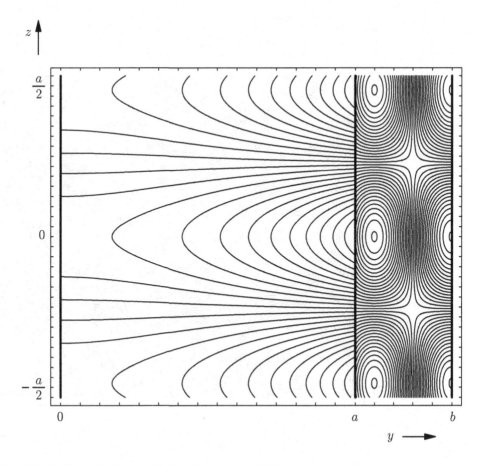

Fig. 6.6–4: Electric lines of force of the second TM-mode

$$\varepsilon_1/\varepsilon_2 = 0.2; \ \omega = 6 \cdot 10^{10} \ [\text{s}^{-1}]; \ \beta_{TM} = 215.3 \ [\text{m}^{-1}]$$

This approach already satisfies the first boundary condition and the two other boundary conditions require

$$\left.\frac{\partial \underline{A}_{TM}}{\partial y}\right|_{\substack{y<a \\ y\to a}} = \left.\frac{\partial \underline{A}_{TM}}{\partial y}\right|_{\substack{y>a \\ y\to a}} ; \qquad \left.\frac{\beta_1^2 - \beta_{TM}^2}{\varepsilon_1}\underline{A}_{TM}\right|_{\substack{y<a \\ y\to a}} = \left.\frac{\beta_2^2 - \beta_{TM}^2}{\varepsilon_2}\underline{A}_{TM}\right|_{\substack{y>a \\ y\to a}} .$$

The evaluation of the boundary conditions leads to

$$\frac{\underline{C}_1}{\underline{C}_2}\sqrt{\frac{\beta_1^2 - \beta_{TM}^2}{\beta_2^2 - \beta_{TM}^2}} = \frac{\cos(\sqrt{\beta_2^2 - \beta_{TM}^2}(a - b))}{\cos(\sqrt{\beta_1^2 - \beta_{TM}^2}a)}$$

$$\frac{\varepsilon_2}{\varepsilon_1}\frac{\underline{C}_1}{\underline{C}_2}\frac{(\beta_1^2 - \beta_{TM}^2)}{(\beta_2^2 - \beta_{TM}^2)} = \frac{\sin(\sqrt{\beta_2^2 - \beta_{TM}^2}(a - b))}{\sin(\sqrt{\beta_1^2 - \beta_{TM}^2}a)}$$

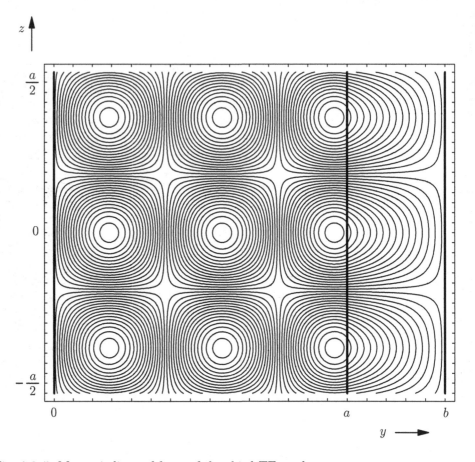

Fig. 6.6–5: Magnetic lines of force of the third TE-mode

$$\varepsilon_1/\varepsilon_2 = 2; \; \omega = 8.2 \cdot 10^{10} \; [\text{s}^{-1}]; \; \beta_{TE} = 273.5 \; [\text{m}^{-1}]$$

and gives rise for the eigenvalue equation

$$\sqrt{\beta_1^2 - \beta_{TM}^2} \, \tan(\sqrt{\beta_1^2 - \beta_{TM}^2}\, a) \; = \; \frac{\varepsilon_1}{\varepsilon_2} \sqrt{\beta_2^2 - \beta_{TM}^2} \, \tan(\sqrt{\beta_2^2 - \beta_{TM}^2}\,(a - b)),$$

for the determination of the phase constants $\beta_{TM}$ of the TM-modes. As the eigenvalue equation has to be solved numerically, it is useful to apply the substitutions $c^2 = a^2(\beta_1^2 - \beta_2^2)$ and $u^2 = a^2(\beta_1^2 - (\beta_{TM})^2)$, with the result:

$$u \tan(u) = \varepsilon_1/\varepsilon_2 \sqrt{c^2 - u^2} \, \tanh\left[\sqrt{c^2 - u^2}(b/a - 1)\right].$$

An analog approach holds for the TE-modes. With

$$\frac{\partial^2 \underline{F}_{TE}}{\partial y^2} + (\beta_{1,2}^2 - \beta_{TE}^2)\,\underline{F}_{TE} = 0 \; ; \qquad \frac{\partial \underline{F}_{TE}}{\partial x} = 0$$

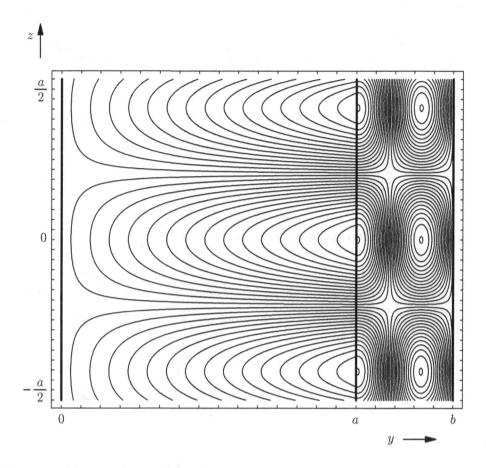

Fig. 6.6–6: Magnetic lines of force of the second TE-mode
$$\varepsilon_1/\varepsilon_2 = 0.2; \ \omega = 7.17 \cdot 10^{10} \ [\text{s}^{-1}]; \ \beta_{TE} = 239.2 \ [\text{m}^{-1}]$$

the solution takes the form

$$\underline{F}_{TE} = \left\{ \begin{array}{l} \underline{C}_3 \cos(\sqrt{\beta_1^2 - \beta_{TE}^2}\,y) \\[2mm] \underline{C}_4 \cos(\sqrt{\beta_2^2 - \beta_{TE}^2}(y-b)) \end{array} \right\} \exp(\mp j\beta_{TE}z); \qquad \begin{array}{l} 0 < y < a \\[2mm] a < y < b, \end{array}$$

that satisfies the boundary condition $(\partial \underline{F}_{TE}/\partial y)|_{y=0,b} = 0$.

The evaluation of the two other boundary conditions in $x = a$ results in

$$\frac{\varepsilon_2}{\varepsilon_1} \frac{\underline{C}_3}{\underline{C}_4} \sqrt{\frac{\beta_1^2 - \beta_{TE}^2}{\beta_2^2 - \beta_{TE}^2}} = \frac{\sin(\sqrt{\beta_2^2 - \beta_{TE}^2}(a-b))}{\sin(\sqrt{\beta_1^2 - \beta_{TE}^2}\,a)}$$

and

$$\frac{\varepsilon_2}{\varepsilon_1}\frac{\underline{C}_3}{\underline{C}_4}\frac{\beta_1^2 - \beta_{TE}^2}{\beta_2^2 - \beta_{TE}^2} = \frac{\cos(\sqrt{\beta_2^2 - \beta_{TE}^2}(a - b))}{\cos(\sqrt{\beta_1^2 - \beta_{TE}^2}a)},$$

and leads to the eigenvalue equation

$$\sqrt{\beta_1^2 - \beta_{TE}^2}\cot(\sqrt{\beta_1^2 - \beta_{TE}^2}a) = \sqrt{\beta_2^2 - \beta_{TE}^2}\cot(\sqrt{\beta_2^2 - \beta_{TE}^2}(a - b)),$$

out of which the phase constants $\beta_{TE}$ of the TE-Modes follow. Again, with the sub stitutions $c^2 = a^2(\beta_1^2 - \beta_2^2)$ and $u^2 = a^2(\beta_1^2 - (\beta_{TE})^2)$ one gets

$$u\cot(u) = -\sqrt{c^2 - u^2}\coth\left[\sqrt{c^2 - u^2}(b/a - 1)\right].$$

## 6.7   Group of Hertzian Dipoles

Four Hertzian dipoles with moments $\vec{e}_z\underline{I}s$ ($s \ll \lambda$) are located at positions $(\pm a; 0; 0$ and $(0; 0; \pm b)$ in the homogeneous space of permittivity $\varepsilon$ and permeability $\mu$.

Calculate the far-field of the group and find the zeros of the radiation pattern in th plane $x = 0$ for $b = 2\lambda$.

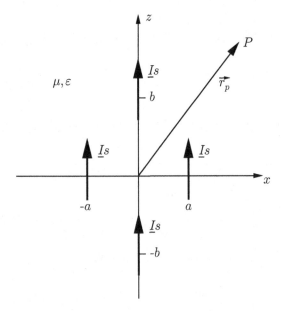

The magnetic far-field $\vec{\underline{H}}_0$ at the observation point $\vec{r}_p$ of a single dipole $\vec{e}_z\underline{I}s$ locate at the point of origin is

$$\vec{\underline{H}}_0 = \frac{j\beta}{4\pi}\underline{I}s\left(\vec{e}_z \times \frac{\vec{r}_p}{r_p}\right)\frac{\exp(-j\beta r_p)}{r_p}; \qquad \beta = \omega\sqrt{\mu\varepsilon} = 2\pi/\lambda.$$

If the same dipole is located at position $\vec{r}_q$, then the excited far field is

$$\vec{\underline{H}} = \frac{j\beta}{4\pi} \underline{I}s \left( \vec{e}_z \times \frac{\vec{r}_p}{r_p} \right) \frac{\exp(-j\beta r_p)}{r_p} \exp(j\beta \vec{r}_q \vec{r}_p / r_p) = \vec{\underline{H}}_0 \exp(j\beta \vec{r}_q \vec{r}_p / r_p) \,.$$

With  $\vec{r}_q = \pm \vec{e}_x a$,  $\vec{r}_q = \pm \vec{e}_z b$,  and

$$\vec{r}_p / r_p = \vec{e}_x \sin\vartheta \cos\varphi + \vec{e}_y \sin\vartheta \sin\varphi + \vec{e}_z \cos\vartheta = \vec{e}_r$$

it follows for the far-field of the group

$$\vec{\underline{H}} = \vec{\underline{H}}_0 \left[ \exp(j\beta a \sin\vartheta \cos\varphi) + \exp(-j\beta a \sin\vartheta \cos\varphi) + \right.$$

$$\left. + \exp(j\beta b \cos\vartheta) + \exp(-j\beta b \cos\vartheta) \right]$$

$$= 2\vec{\underline{H}}_0 \left[ \cos(\beta a \sin\vartheta \cos\varphi) + \cos(\beta b \cos\vartheta) \right] = 2\vec{\underline{H}}_0 f(\vartheta, \varphi) \,.$$

Within the far-field the electric field becomes

$$\vec{\underline{E}} = -Z \vec{e}_r \times \vec{\underline{H}} = -2Z \vec{e}_r \times \vec{\underline{H}}_0 f(\vartheta, \varphi)$$

$$= -Z \frac{j\beta}{2\pi} \underline{I}s \cdot \underbrace{\vec{e}_r \times (\vec{e}_z \times \vec{e}_r)}_{= -\vec{e}_\vartheta \sin\vartheta} \cdot \frac{\exp(-j\beta r_p)}{r_p} f(\vartheta, \varphi)$$

$$\vec{\underline{E}} = \vec{e}_\vartheta \frac{j\beta}{2\pi} \underline{I} Z s \frac{\exp(-j\beta r_p)}{r_p} \sin\vartheta \, f(\vartheta, \varphi) \,,$$

where $Z = \sqrt{\mu/\varepsilon}$ is the wave impedance.  Therewith the time-average $\vec{S}_-$ of the Poynting vector $\vec{S}$ in the far-field is

$$\vec{S}_- = 1/2 \operatorname{Re}\left\{ \vec{\underline{E}} \times \vec{\underline{H}}^* \right\} = -Z/2(\vec{e}_r \times \vec{\underline{H}}) \times \vec{\underline{H}}^* = \vec{e}_r Z/2 |\vec{\underline{H}}|^2 = \vec{e}_r S_-(r_p, \vartheta, \varphi)$$

$$S_-(r_p, \vartheta, \varphi) = 2Z|\vec{\underline{H}}_0|^2 f^2(\vartheta, \varphi) = 2Z(\beta/4\pi)^2 \underline{I}\, \underline{I}^* \, s^2 \sin^2\vartheta \, f^2(\vartheta, \varphi)/r_p^2$$

$$\Rightarrow \quad S_-(r_p, \vartheta, \varphi) = Z/2(|\underline{I}|s/\lambda)^2 \sin^2\vartheta \, f^2(\vartheta, \varphi)/r_p^2 \,.$$

This leads to the radiation pattern

$$\Phi(\vartheta, \varphi) = S_- \, r_p^2 = Z/2(|\underline{I}|s/\lambda)^2 \sin^2\vartheta \, f^2(\vartheta, \varphi) \,.$$

It follows for the plane $x = 0$ with $\varphi = \pm\pi/2$

$$\Phi(\vartheta, \varphi = \pm\pi/2) = Z/2(|\underline{I}|s/\lambda)^2 \sin^2\vartheta \, [1 + \cos(\beta b \cos\vartheta)]^2 \,.$$

Finally with $b = 2\lambda$ the zeros of the radiation pattern are given by

$$\beta b \cos\vartheta = (2k+1)\pi \,; \qquad k = -2; -1; 0, 1$$

$$\cos\vartheta = (2k+1)/4 \,.$$

Furthermore additional zeros exist at $\vartheta = 0$ and $\vartheta = \pi$.

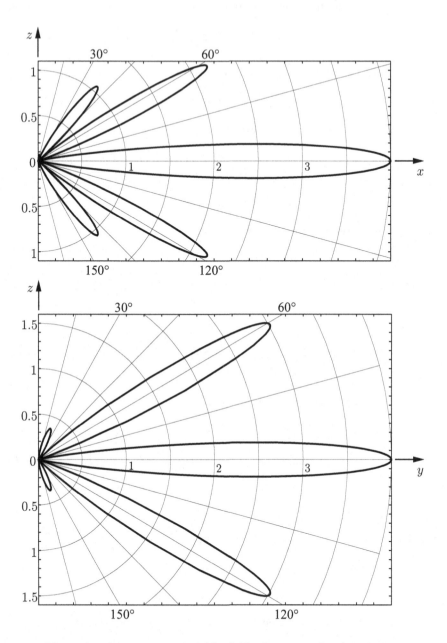

Fig. 6.7–1: Vertical radiation pattern $\Phi(\vartheta, \varphi)/\Phi_0$ for $\varphi = 0$ (top) and
$\varphi = \pi/2$ (bottom) with $\beta b = 4\pi$, $\beta a = 2\pi$, and $\Phi_0 = Z/2(|\underline{I}|s/\lambda)^2$

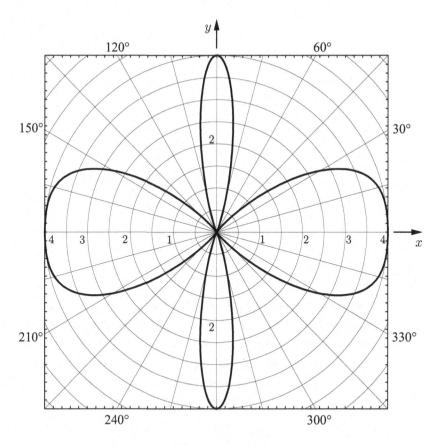

Fig. 6.7–2: Horizontal radiation pattern $\Phi(\vartheta, \varphi)/\Phi_0$ for $\vartheta = \pi/2$ and $\beta a = 2\pi$

## 6.8 Linear Antenna in Front of a Conducting Plane

A linear antenna of length $0 < z < 3/4\lambda$ is positioned in front of a perfect conducting plane in $z = 0$ and carries the current

$$i(z,t) = \mathrm{Re}\left\{i_0 \cos(2\pi z/\lambda)\exp(j\omega t)\right\} ; \qquad 2\pi/\lambda = \beta = \omega\sqrt{\mu\varepsilon}.$$

Calculate the far-field and the radiation pattern in $z > 0$, where the permittivity is $\varepsilon$ and the permeability is $\mu$.

At the perfect conducting boundary the tangential component $\vec{e}_z \times \vec{E}|_{z=0}$ vanishes and it is thus possible to apply the method of images to describe the field in $z > 0$. For this purpose the given current $i(z,t)$ on the $z$-axis is extended to the total length $-3/4\lambda < z < 3/4\lambda$ and the antenna is assumed to be in the homogeneous space with material properties $\mu, \varepsilon$.

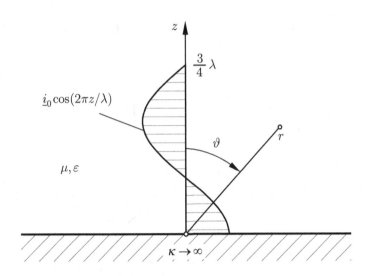

The excited magnetic far-field can be determined by an integration of the contribu-
tions $d\vec{\underline{H}}(\vec{r}_p)$ of infinitesimal dipoles $\vec{e}_z i_0\, dz$ over the total length $-3/4\lambda < z < 3/4\lambda$.
Alternatively the contribution of three $\lambda/2$-dipoles can be summarized. The single
$\lambda/2$-dipole carries the current $i(z,t)$ on the $z$-axis but is restricted to the length
$-\lambda/4 < z < \lambda/4$. Now at first the magnetic far-field of a $\lambda/2$-dipole is calculated
by an integration of the contributions of Hertzian dipoles.

A Hertzian current element $\vec{e}_z i_0 dz_q$ at position $\vec{r}_q = \vec{e}_z\, z_q$ in the homogeneous space
provides the differential contribution $d\vec{\underline{H}}$ to the magnetic far-field

$$d\vec{\underline{H}}(\vec{r}_p) = j\frac{\beta}{4\pi} i_0\, dz_q \left(\vec{e}_z \times \frac{\vec{r}_p}{r_p}\right) \frac{\exp(-j\beta r_p)}{r_p} \exp(j\beta z_q \vec{e}_z \vec{r}_p / r_p).$$

According to the specified current in $-\lambda/4 < z < \lambda/4$ ($\lambda/2$−dipole) this leads th
following magnetic far-field $\vec{\underline{H}}_D$.

$$\vec{\underline{H}}_D(\vec{r}_p) = j\frac{\beta}{4\pi} i_0 \left(\vec{e}_z \times \frac{\vec{r}_p}{r_p}\right) \frac{\exp(-j\beta r_p)}{r_p} \int\limits_{-\lambda/4}^{\lambda/4} \cos(\beta z_q)\, \exp(j\beta z_q \cos\vartheta)\, dz_q$$

$$= \vec{e}_\varphi\, j\frac{i_0}{2\pi} \frac{\exp(-j\beta r_p)}{r_p} \frac{\cos(\pi/2 \cos\vartheta)}{\sin\vartheta} = \vec{e}_\varphi \underline{H}_D(r_p, \vartheta)$$

The resulting field $\vec{\underline{H}}_G$ of the antenna with the current $i(z,t)$ in $-3/4\lambda < z < 3/4\lambda$
is the superposition of the fields of three $\lambda/2$-dipoles with midpoints at $z_q = 0, \pm\lambda/2$
taking into account the correct phase of the currents.

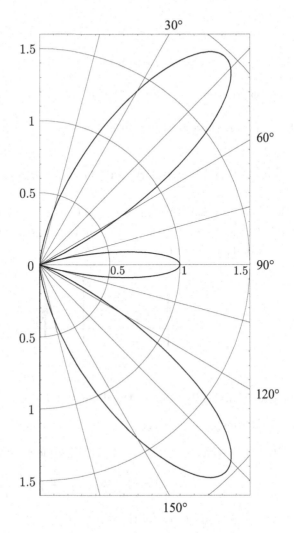

Fig. 6.8–1: Vertical radiation pattern $\Phi(\vartheta)/\Phi_0$, with $\Phi_0 = |\underline{i}_0|^2\, Z/(8\pi^2)$

$$\underline{\vec{H}}_G = \vec{e}_\varphi \underline{H}_D\left[1 - \exp(j\beta\lambda/2\cos\vartheta) - \exp(-j\beta\lambda/2\cos\vartheta)\right] = \vec{e}_\varphi \underline{H}_D\left[1 - 2\cos(\pi\cos\vartheta)\right]$$

With it the related electric far-field is

$$\underline{\vec{E}}_G = Z(\underline{\vec{H}}_G \times \vec{e}_r)\,; \qquad \vec{e}_r = \vec{r}_p/r_p\,; \qquad Z = \sqrt{\mu/\varepsilon}\,.$$

Finally the time-averaged Poynting vector $\vec{S}$ is

$$\vec{S}_- = \frac{1}{2}\mathrm{Re}\left\{\underline{\vec{E}}_G \times \underline{\vec{H}}_G^*\right\} = \frac{1}{2}\,Z|\underline{\vec{H}}_G|^2\,\vec{e}_r = \vec{e}_r\, S_-$$

$$S_- = \frac{Z}{2} \frac{|\underline{i}_0|^2}{4\pi^2} \frac{1}{r_p^2} \frac{\cos^2(\pi/2\cos\vartheta)}{\sin^2\vartheta} [1 - 2\cos(\pi\cos\vartheta)]^2$$

and therewith the radiation pattern takes the form

$$\Phi(\vartheta) = S_- r_p^2 = |\underline{i}_0|^2 \frac{Z}{8\pi^2} \left[ \frac{\cos(\pi/2\cos\vartheta)(1 - 2\cos(\pi\cos\vartheta))}{\sin\vartheta} \right]^2 .$$

The horizontal radiation pattern in $\vartheta = \pi/2$ is a circle.

## 6.9 Hertzian Dipoles Along the $x$-Axis

Four Hertzian dipoles with moments $\vec{e}_z I s$ ($s \ll \lambda$) are located at positions
$(-c \pm a, 0, 0)$ and $(c \pm b, 0, 0)$ with $c > a, b$ in the homogeneous space of permittivity $\varepsilon$
and permeability $\mu$. Find the magnetic far-field.

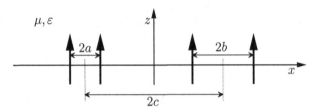

The magnetic far-field $\vec{\underline{H}}_1(\vec{r}_p)$ of the dipoles at $(-c \pm a, 0, 0)$ alone is

$$\vec{\underline{H}}_1(\vec{r}_p) = \vec{\underline{H}}_0 [\exp(j\beta a \sin\vartheta \cos\varphi) + \exp(-j\beta a \sin\vartheta \cos\varphi)] = 2\vec{\underline{H}}_0 \cos(\beta a \sin\vartheta \cos\varphi)$$

where $\vec{\underline{H}}_0$ is the magnetic far-field of one dipole located at the point of origin

$$\vec{\underline{H}}_0(\vec{r}_p) = \frac{j\beta}{4\pi} I s \left( \vec{e}_z \times \frac{\vec{r}_p}{r_p} \right) \frac{\exp(-j\beta r_p)}{r_p} .$$

Similarly the far-field $\vec{\underline{H}}_2(\vec{r}_p)$ of the dipoles at $(c \pm b, 0, 0)$ alone is

$$\vec{\underline{H}}_2(\vec{r}_p) = 2\vec{\underline{H}}_0 \cos(\beta b \sin\vartheta \cos\varphi) .$$

Thus, the magnetic far-field of only two dipoles is independent of $c$.

The resulting far-field $\vec{\underline{H}}(\vec{r}_p)$ of the four dipoles is given by

$$\vec{\underline{H}}(\vec{r}_p) = \vec{\underline{H}}_1 \exp(-j\beta c \sin\vartheta \cos\varphi) + \vec{\underline{H}}_2 \exp(j\beta c \sin\vartheta \cos\varphi)$$
$$= \left( \vec{\underline{H}}_1 + \vec{\underline{H}}_2 \right) \cos(\beta c \sin\vartheta \cos\varphi) + j \left( \vec{\underline{H}}_2 - \vec{\underline{H}}_1 \right) \sin(\beta c \sin\vartheta \cos\varphi) .$$

If both elementary groups are identical ($a = b \Rightarrow \vec{\underline{H}}_1 = \vec{\underline{H}}_2$) one gets the simple
solution

$$\vec{\underline{H}}(\vec{r}_p) = 4\vec{\underline{H}}_0 \cos(\beta a \sin\vartheta \cos\varphi) \cos(\beta c \sin\vartheta \cos\varphi) .$$

## 6.10   Radiation Patterns of Antenna Arrays

The magnetic far-field of an arbitrary system of linear antennas or dipoles is always of the form

$$\underline{\vec{H}}_0(\vec{r}_p) = \underline{\vec{H}}_E(\vec{r}_p/r_p)\,\frac{r_0}{r_p}\,\exp(-j\beta r_p)\,.$$

where $r_0$ is an arbitrary constant and $\underline{\vec{H}}_E$ describes the specific field of the source. Thus, related to the far-field the system of antennas behaves like a point source in the origin. The radiation pattern is given by

$$\Phi_0(\vartheta,\varphi) = \frac{1}{2}Z|\underline{\vec{H}}_0(\vec{r}_p)|^2 r_p^2 = \frac{1}{2}Z|\underline{\vec{H}}_E(\vec{r}_p/r_p)|^2 r_0^2\,.$$

Find the magnetic far-field and the radiation pattern when four systems are positioned equidistantly on the $x$-axis. The distance between two adjacent systems is $2a$.

For the calculation of the far-field of multiple identical antennas applies

$$\underline{\vec{H}}(\vec{r}_p) = \sum_i \underline{\vec{H}}_0(\vec{r}_p)\,\exp(j\beta\vec{r}_{qi}\vec{r}_p/r_p)\,,$$

where $\vec{r}_{qi}$ points to the position of the $i$-th antenna. Thus for the present problem we get

$$\underline{\vec{H}}(\vec{r}_p) = \underline{\vec{H}}_0(\vec{r}_p)\Big[\exp(j\beta\vec{e}_x a\vec{r}_p/r_p) + \exp(-j\beta\vec{e}_x a\vec{r}_p/r_p)+$$
$$+ \exp(j\beta\vec{e}_x 3a\vec{r}_p/r_p) + \exp(-j\beta\vec{e}_x 3a\vec{r}_p/r_p)\Big]\,.$$

With $\vec{e}_x\vec{r}_p/r_p = c = \sin\vartheta\,\cos\varphi$ it follows

$$\underline{\vec{H}}(\vec{r}_p) = \underline{\vec{H}}_0(\vec{r}_p)\Big[\exp(j\beta ac) + \exp(-j\beta ac) + \exp(j\beta 3ac) + \exp(-j\beta 3ac)\Big]$$
$$= \underline{\vec{H}}_0(\vec{r}_p)\Big[2\cos(\beta ac) + 2\cos(\beta 3ac)\Big]\,.$$

The corresponding radiation pattern can be expressed by

$$\Phi(\vartheta,\varphi) = \Phi_0(\vartheta,\varphi)\Phi_{G4}(\vartheta,\varphi)$$

where

$$\Phi_{G4}(\vartheta,\varphi) = \Big[2\cos(\beta ac) + 2\cos(\beta 3ac)\Big]^2$$

is called array factor of the group with four antennas.

An alternative description of the magnetic field follows from the approach of problem 6.9. In the first step only two antennas are combined

$$\underline{\vec{H}}_1(\vec{r}_p) \;=\; \underline{\vec{H}}_0(\vec{r}_p)\Big[\exp(j\beta ac) + \exp(-j\beta ac)\Big] \;=\; \underline{\vec{H}}_0(\vec{r}_p)\,2\cos(\beta ac)\,.$$

Finally the new combination of two combined antennas with distance $4a$ leads to the
far-field

$$\underline{\vec{H}}(\vec{r}_p) \;=\; \underline{\vec{H}}_1(\vec{r}_p)\Big[\exp(j2\beta ac) + \exp(-j2\beta ac)\Big] \;=\; \underline{\vec{H}}_1(\vec{r}_p)\,2\cos(2\beta ac)$$

$$=\; \underline{\vec{H}}_0(\vec{r}_p)\,2\cos(2\beta ac)\,2\cos(\beta ac)\,.$$

Now the array factor is

$$\Phi_{G4}(\vartheta,\varphi) \;=\; \Phi_{G2}(\vartheta,\varphi,2a)\Phi_{G2}(\vartheta,\varphi,4a) \;=\; \Big[4\cos(\beta ac)\cos(\beta 2ac)\Big]^2$$

where

$$\Phi_{G2}(\vartheta,\varphi,d) \;=\; 4\cos^2(1/2\beta dc)$$

is the array factor of two antennas with distance $d$.

Fig. 6.10–1: Array with eight antennas and array factor
$$\Phi_{G8}(\vartheta,\varphi) = \Phi_{G2}(\vartheta,\varphi,a)\Phi_{G2}(\vartheta,\varphi,2a)\Phi_{G2}(\vartheta,\varphi,4a)$$

## 6.11   Waveguide with Sections of Different Dielectrics

A rectangular waveguide with perfectly conducting boundaries and cross sectional dimensions $a$ and $b < a$ is of infinite length in $z$-direction. In $z < 0$ a dielectric material
with permittivity $\varepsilon_1$ and in $z > 0$ material with permittivity $\varepsilon_2 < \varepsilon_1$ is inserted. The
permeability $\mu$ is constant.

Find the function, that specifies the dependence on the $z$-coordinate of the Mode
in $z > 0$, when in $z < 0$ the fundamental mode is excited, propagating in positive
$z$-direction at the frequency

$$\omega \;=\; 1/2\,(\omega_{c1} + \omega_{c2}) \;=\; 1/2\,\omega_{c1}(1 + \sqrt{\varepsilon_1/\varepsilon_2})\,; \qquad \omega_{c1,2} = \pi/(a\sqrt{\mu\varepsilon_{1,2}})\,.$$

The phase constants $\beta_{z_{mn}}^{(1)}$ of the modes in $z < 0$ are

$$\beta_{z_{mn}}^{(1)} = \sqrt{\beta_1^2 - ((m\pi/a)^2 + (n\pi/b)^2)} = \sqrt{\beta_1^2 - \beta_{mn}^2} \; ; \qquad \beta_1^2 = \omega^2 \mu \varepsilon_1 \, .$$

For the fundamental mode holds $m = 1$ and $n = 0$ because of $a > b$ and thus the cut-off frequency $\omega_{c1}$ is

$$\beta_{z_{10}}^{(1)} = 0 \, ; \qquad \beta_1^2 = \beta_{10}^2 = (\pi/a)^2 = \omega_{c1}^2 \, \mu \varepsilon_1 \, ; \qquad \omega_{c1} = \pi/(a\sqrt{\mu \varepsilon_1}) \, .$$

Hence the excitation of the waves in $z < 0$ occurs at a frequency $\omega$ greater than the cut-off limit $\omega_{c1}$, as long as $\varepsilon_1 > \varepsilon_2$.

In $z > 0$ the phase constants follow from

$$\beta_{z_{mn}}^{(2)} = \sqrt{\beta_2^2 - \beta_{mn}^2} \; ; \qquad \beta_2^2 = \omega^2 \, \mu \varepsilon_2$$

$$\beta_{z_{mn}}^{(2)} = \sqrt{1/4 \, \omega_{c1}^2 \left(1 + \sqrt{\varepsilon_1/\varepsilon_2}\right)^2 \mu \varepsilon_2 - \beta_{mn}^2} \, .$$

For the fundamental mode $\beta_{z_{10}}^{(2)}$ applies with $\beta_{10} = \pi/a = \omega_{c1} \sqrt{\mu \varepsilon_1}$

$$\beta_{z_{10}}^{(2)} = \sqrt{1/4 \, \omega_{c1}^2 (1 + \sqrt{\varepsilon_1/\varepsilon_2})^2 \, \mu \varepsilon_2 - \omega_{c1}^2 \, \mu \varepsilon_1}$$

$$= \omega_{c1} \sqrt{\mu \varepsilon_1 \left[1/4(1 + \sqrt{\varepsilon_1/\varepsilon_2})^2 \, \varepsilon_2/\varepsilon_1 - 1\right]}$$

$$\beta_{z_{10}}^{(2)} = \pi/a \sqrt{1/4 \, \varepsilon_2/\varepsilon_1 + 1/2 \sqrt{\varepsilon_2/\varepsilon_1} - 3/4} \, .$$

Now the radicand is negative because of $\varepsilon_2/\varepsilon_1 < 1$ and thus in $z > 0$ the fundamental mode and all other modes decay exponentially with increasing distance $z$. For the fundamental mode holds

$$\exp\left(-\pi/a \sqrt{3/4 - 1/4 \, \varepsilon_2/\varepsilon_1 - 1/2\sqrt{\varepsilon_2/\varepsilon_1}} \, z\right) \, .$$

All higher order modes decay with the function

$$\exp\left(-\sqrt{m^2 + (na/b)^2 - 1/4 \, \varepsilon_2/\varepsilon_1(1 + \sqrt{\varepsilon_1/\varepsilon_2})^2} \, \pi \, z/a\right) \, .$$

## 6.12   Reflection of a Plane Wave at a Conducting Half-Plane

A plane wave with the electric field

$$\vec{E}(z,t) = \mathrm{Re}\left\{\vec{e}_x\, E_0\, \exp(-j\beta_0 z)\, \exp(j\omega t)\right\}\,; \qquad \beta_0 = \omega\sqrt{\mu_0\varepsilon_0}$$

impinges on a slab $0 \le z < d$ with conductivity $\kappa$, permittivity $\varepsilon$, and permeability $\mu$. The backside of the slab $z \ge d$ is perfectly conducting and in $z < 0$ the permittivity and permeability are $\varepsilon_0$ and $\mu_0$.

Calculate the field in $z \le d$ and the current sheet in $z = d$. Use your solutions to analyze the limits $\kappa = 0$, $\mu = \mu_0$, and $\varepsilon = \varepsilon_0$.

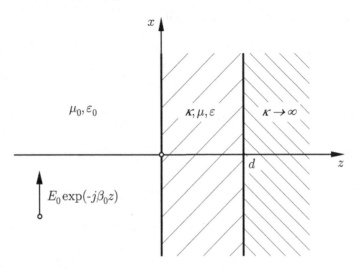

The approach for the complex amplitude of the electric field $\underline{\vec{E}} = \vec{e}_x\, \underline{E}(z)$ in $z \le d$ is

$$\underline{E}(z) = \begin{cases} E_0 \exp(-j\beta_0 z) + \underline{E}_{02} \exp(j\beta_0 z) & ;\quad z \le 0 \\ \underline{E}_{11} \exp(-j\gamma z) + \underline{E}_{12} \exp(j\gamma z) & ;\quad 0 \le z \le d \end{cases}$$

$$\vec{\gamma} = \vec{e}_z\, \gamma\,; \quad \gamma^2 = (\gamma_r - j\gamma_i)^2 = \beta^2 - \alpha^2\,; \quad \beta^2 = \omega^2\mu\varepsilon\,; \quad \alpha^2 = j\omega\kappa\mu$$

$$\gamma_r = \beta\sqrt{1/2(\sqrt{1 + (\kappa/(\omega\varepsilon))^2} + 1)}\,; \quad \gamma_i = \beta\sqrt{1/2(\sqrt{1 + (\kappa/(\omega\varepsilon))^2} - 1)}\,.$$

For the complex amplitude of the magnetic flux density $\underline{\vec{B}} = \mu\underline{\vec{H}}$ applies $\underline{\vec{B}} = \pm 1/\omega\, \vec{\gamma} \times \underline{\vec{E}}$, where the leading sign denotes waves propagating in positive or negative $z$-direction.

Consequently, the complex amplitude $\vec{H} = \vec{e}_y \underline{H}(z)$ is

$$\underline{H}(z) = \begin{cases} 1/Z_0(E_0 \exp(-j\beta_0 z) - \underline{E}_{02} \exp(j\beta_0 z)) & ; \quad z \leq 0 \\ \gamma/(\omega\mu)(\underline{E}_{11} \exp(-j\gamma z) - \underline{E}_{12} \exp(j\gamma z)) & ; \quad 0 \leq z \leq d \end{cases}$$

with $Z_0 = \sqrt{\mu_0/\varepsilon_0}$.

The unknown complex amplitudes $\underline{E}_{02}, \underline{E}_{11}$, and $\underline{E}_{12}$ follow from the boundary conditions in $z = 0$ and $z = d$.

$$\vec{e}_z \times \vec{E}\Big|_{z=d} = 0 ; \quad \vec{e}_z \times \vec{E}\Big|_{\substack{z<0 \\ z\to 0}} = \vec{e}_z \times \vec{E}\Big|_{\substack{z>0 \\ z\to 0}} ; \quad \vec{e}_z \times \vec{H}\Big|_{\substack{z<0 \\ z\to 0}} = \vec{e}_z \times \vec{H}\Big|_{\substack{z>0 \\ z\to 0}}$$

$$\underline{E}_{11} \exp(-j\gamma d) + \underline{E}_{12} \exp(j\gamma d) = 0 \quad \Rightarrow \quad \underline{E}_{12} = -\underline{E}_{11} \exp(-2j\gamma d)$$

$$E_0 + \underline{E}_{02} = \underline{E}_{11} + \underline{E}_{12} ; \quad 1/Z_0(E_0 - \underline{E}_{02}) = \gamma/(\omega\mu)(\underline{E}_{11} - \underline{E}_{12})$$

$$\frac{1 - \underline{E}_{02}/E_0}{1 + \underline{E}_{02}/E_0} = \frac{1-r}{1+r} = \frac{\gamma Z_0}{\omega\mu} \frac{1 + \exp(-2j\gamma d)}{1 - \exp(-2j\gamma d)} = \frac{\gamma Z_0}{j\omega\mu} \frac{\cos(\gamma d)}{\sin(\gamma d)} = q$$

$$\underline{E}_{02} = r E_0 ; \quad \underline{E}_{11} = E_0 \frac{1+r}{1 - \exp(-2j\gamma d)}$$

$$\underline{E}_{12} = -E_0(1+r) \frac{\exp(-2j\gamma d)}{1 - \exp(-2j\gamma d)} ; \quad r = (1-q)/(1+q)$$

The current sheet $\vec{K} = -\vec{e}_z \times \vec{H}\Big|_{z=d} = \vec{e}_x \underline{K}$ is

$$\underline{K} = \gamma/(\omega\mu)(\underline{E}_{11} \exp(-j\gamma d) - \underline{E}_{12} \exp(j\gamma d)) = -j E_0 \frac{1+r}{\sin(\gamma d)} \frac{\gamma}{\omega\mu} .$$

With $\kappa = 0$ it is $\gamma_i = 0$ and $\gamma = \beta$ and furthermore

$$q = \frac{Z_0}{jZ} \frac{\cos(\beta d)}{\sin(\beta d)} ; \quad Z = \sqrt{\mu/\varepsilon} .$$

If, in addition, the limits $\mu = \mu_0$ and $\varepsilon = \varepsilon_0$ are built, then with $Z = Z_0$ and $\beta = \beta_0$ the reflection factor is $r = -\exp(-2j\beta_0 d)$. The complex amplitudes are

$$\underline{E}_{11} = E_0 ; \quad \underline{E}_{12} = \underline{E}_{02} = -E_0 \exp(-2j\beta_0 d) ; \quad \underline{K} = 2E_0/Z \exp(-j\beta_0 d) .$$

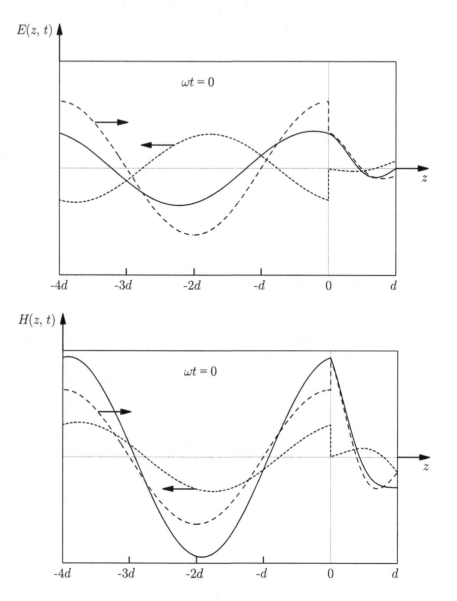

Fig. 6.12–1: Graphs of the electric and magnetic field as a function of the $z$-coordinate at the time $t = 0$. Additionally the partial waves propagating in different directions are plotted separately.

$\mu = \mu_0; \quad \varepsilon = \varepsilon_0; \quad \kappa = 2 \; [\text{S/m}]; \quad d/|\gamma| = 2.42 \cdot 10^{-5}$

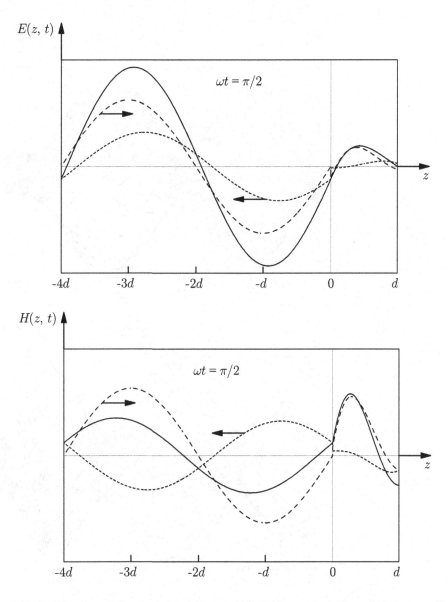

Fig. 6.12–2: Graphs of the electric and magnetic field as a function of the $z$-coordinate at the time $\omega t = \pi/2$. Additionally the partial waves propagating in different directions are plotted separately. Parameters: see Fig. 6.12–1

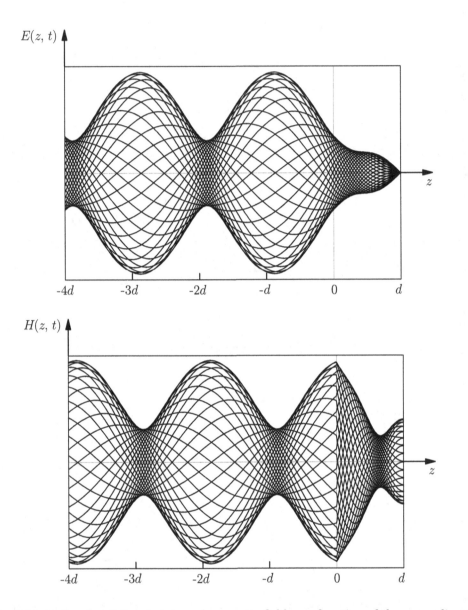

Fig. 6.12–3: Graphs of the electric and magnetic field as a function of the $z$-coordinat
at various times $t$. Parameters: see Fig. 6.12–1

## 6.13 Guided Waves in a Dielectric Slab Waveguide

A dielectric slab $a < x < b$ with permittivity $\varepsilon_1$ is positioned beside a perfectly conducting wall $x \leq 0$. Outside of the slab the permittivity is $\varepsilon_0$ and the permeability $\mu$ is constant. The slab is capable to guide both transverse magnetic (TM-) and transverse electric (TE-)waves in the $z$-direction.

Find the eigenvalue equations for the determination of the phase constants of the modes, that are independent of the $y$-coordinate.

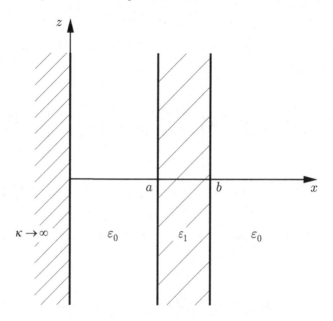

### TM-Waves (-Modes)

These modes can be derived from the vector potential $\vec{\underline{A}}_{TM} = \vec{e}_z \, \underline{A}_{TM}(x, z)$ with

$$\vec{\underline{B}}_{TM} = \text{rot}\,[\vec{e}_z \, \underline{A}_{TM}(x, z)] = \mu \vec{\underline{H}}_{TM} = \vec{e}_y \, \mu \underline{H}_{TM}(x, z) = -\vec{e}_y \, \partial \underline{A}_{TM}/\partial x \,.$$

Obviously the magnetic field has only one component, that satisfies the differential equation

$$\Delta \underline{H}_{TM} + \beta_{0,1}^2 \, \underline{H}_{TM} = \frac{\partial^2 \underline{H}_{TM}}{\partial x^2} + \frac{\partial^2 \underline{H}_{TM}}{\partial z^2} + \beta_{0,1}^2 \underline{H}_{TM} = 0; \quad \beta_{0,1}^2 = \omega^2 \mu \varepsilon_{0,1} \,.$$

An approach for $\underline{H}_{TM}$, that describes wave propagation in $\pm z$-direction with the phase constant $\beta_z^{TM}$, is

$$\underline{H}_{TM}(x, z) = \underline{H}_0 \underline{X}_{1,2}(x) \, \exp(\mp j \beta_z^{TM} z)$$

$$\frac{d^2 X_1}{dx^2} + \underbrace{(\beta_0^2 - \beta_z^{TM^2})}_{-\alpha^2} X_1 = 0 ; \qquad 0 < x < a; \quad x > b$$

$$\frac{d^2 X_2}{dx^2} + \underbrace{(\beta_1^2 - \beta_z^{TM^2})}_{\beta_x^2 > 0} X_2 = 0 ; \qquad a < x < b .$$

The phase constants $\beta_z^{TM}$ of each mode are of the same value in any region of the half-space $x > 0$ and lie within an interval defined by the material parameters $\varepsilon_0$ and $\varepsilon_1$.

$$\beta_0 < \beta_z^{TM} < \beta_1 ; \qquad \alpha^2 = \beta_z^{TM^2} - \beta_0^2 > 0 ; \qquad \beta_x^2 = \beta_1^2 - \beta_z^{TM^2} > 0$$

Hence the complex amplitude $\underline{H}_{TM}(x,z)$ takes the form

$$\underline{H}_{TM}(x,z) = \underline{H}_0 \left\{ \begin{array}{l} \underline{C}_1 \cosh(\alpha x)/\cosh(\alpha a) \\[2mm] \underline{C}_2 \sin(\beta_x(x-a)) + \\ \quad + \underline{C}_3 \cos(\beta_x(x-a)) \\[2mm] \underline{C}_4 \exp(-\alpha(x-b)) \end{array} \right\} \exp(\mp j\beta_z^{TM} z) ; \quad \begin{array}{l} 0 \le x \le a \\[2mm] a \le x \le b \\[4mm] x \ge b \end{array}$$

$$\Rightarrow \quad \vec{\underline{E}}_{TM} = \frac{1}{j\omega\varepsilon} \operatorname{rot} \vec{\underline{H}}_{TM} = \frac{1}{j\omega\varepsilon}\left[ -\vec{e}_x \frac{\partial \underline{H}_{TM}}{\partial z} + \vec{e}_z \frac{\partial \underline{H}_{TM}}{\partial x} \right].$$

The field must satisfy the boundary conditions

$$\vec{e}_z \vec{\underline{E}}_{TM}\Big|_{x=0} = 0 ; \qquad \underline{H}_{TM}\Big|_{\substack{0 \le x \le a \\ x \to a}} = \underline{H}_{TM}\Big|_{\substack{a \le x \le b \\ x \to a}}$$

$$\underline{H}_{TM}\Big|_{\substack{a \le x \le b \\ x \to b}} = \underline{H}_{TM}\Big|_{\substack{x \ge b \\ x \to b}} ; \qquad \vec{e}_z \vec{\underline{E}}_{TM}\Big|_{\substack{0 \le x \le a \\ x \to a}} = \vec{e}_z \vec{\underline{E}}_{TM}\Big|_{\substack{a \le x \le b \\ x \to a}}$$

$$\vec{e}_z \vec{\underline{E}}_{TM}\Big|_{\substack{a \le x \le b \\ x \to b}} = \vec{e}_z \vec{\underline{E}}_{TM}\Big|_{\substack{x \ge b \\ x \to b}} .$$

The first one is already satisfied by the given approach and the other ones lead to the following equations.

$$\underline{C}_1 = \underline{C}_3 ; \qquad \underline{C}_2 \sin(\beta_x(b-a)) + \underline{C}_3 \cos(\beta_x(b-a)) = \underline{C}_4$$

$$\frac{\alpha}{\varepsilon_0}\underline{C}_1 \tanh(\alpha a) = \frac{\beta_x}{\varepsilon_1}\underline{C}_2$$

$$\frac{\beta_x}{\varepsilon_1}\left[\underline{C}_2 \cos(\beta_x(b-a)) - \underline{C}_3 \sin(\beta_x(b-a))\right] = \frac{-1}{\varepsilon_0}\alpha\underline{C}_4$$

After a straightforward calculation the eigenvalue equation becomes

$$\frac{1 + \tanh(\alpha a)}{(\varepsilon_0/\varepsilon_1 \beta_x/\alpha)^2 - \tanh(\alpha a)} = \frac{\alpha}{\beta_x}\frac{\varepsilon_1}{\varepsilon_0} \tan(\beta_x(b-a))$$

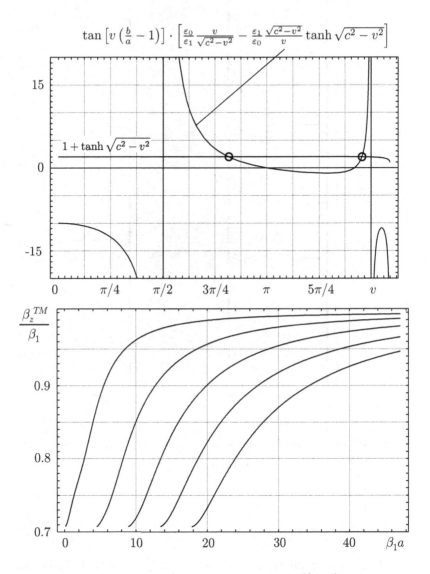

Fig. 6.13–1: Determination of the eigenvalues at $\omega = 15 \cdot 10^{10}$ [s$^{-1}$] and phase constants as a function of the frequency for TM-modes with $\varepsilon_1/\varepsilon_0 = 2$ and $b/a = 2$
$$v^2 = a^2(\beta_1^2 - (\beta_z^{TM})^2); \quad c^2 = a^2(\beta_1^2 - \beta_0^2)$$

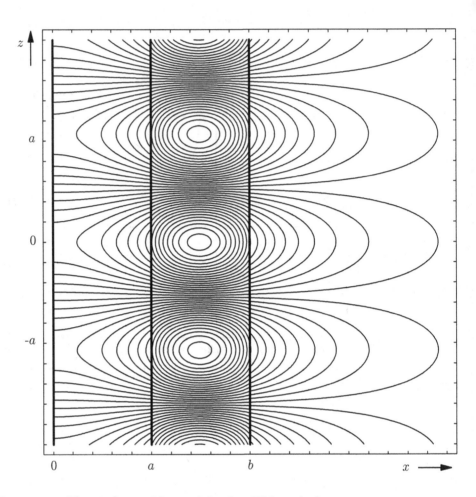

Fig. 6.13–2: Electric lines of force of the first TM-mode for
$$\varepsilon_1/\varepsilon_0 = 2, \ b/a = 2, \ a = 1 [\text{cm}], \text{ and } \omega = 7.5 \cdot 10^{10} \ [\text{s}^{-1}]$$

with

$$(\alpha a)^2 + (\beta_x a)^2 = (\beta_1 a)^2 - (\beta_0 a)^2 ; \qquad \beta_z^{TM} = \sqrt{\beta_1^2 - \beta_x^2} = \sqrt{\alpha^2 + \beta_0^2} .$$

The limited number of solutions for $\beta_x, \beta_z^{TM}, \alpha$ have to be determined numerically.

With $v^2 = a^2(\beta_1^2 - (\beta_z^{TM})^2)$ and $c^2 = a^2(\beta_1^2 - \beta_0^2)$ one gets

$$1 + \tanh \sqrt{c^2 + v^2} =$$

$$= \tan \left[ v \left( \frac{b}{a} - 1 \right) \right] \cdot \left[ \frac{\varepsilon_0}{\varepsilon_1} \frac{v}{\sqrt{c^2 - v^2}} - \frac{\varepsilon_1}{\varepsilon_0} \frac{\sqrt{c^2 - v^2}}{v} \tanh \sqrt{c^2 - v^2} \right] .$$

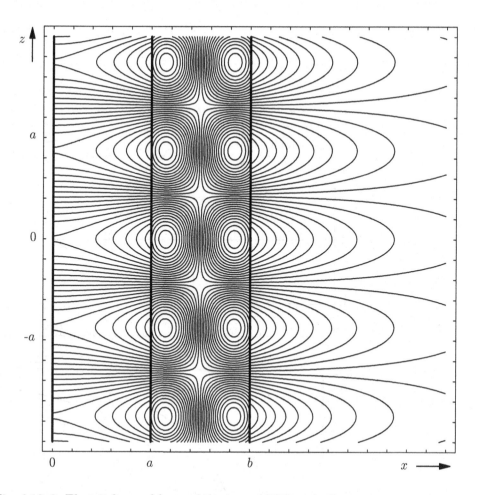

Fig. 6.13–3: Electric lines of force of the second TM-mode for
$\varepsilon_1/\varepsilon_0 = 3$, $b/a = 2$, $a = 1$[cm], and $\omega = 10^{11}$ [s$^{-1}$]

## TE-Waves (-Modes)

The vector potential of the field is $\vec{\underline{F}}_{TE} = \vec{e}_z\,\underline{F}_{TE}(x,z)$ with

$$\vec{\underline{E}}_{TE} = -1/\varepsilon\,\mathrm{rot}\,\vec{\underline{F}}_{TE} = \vec{e}_y\,\underline{E}_{TE}(x,z) = \vec{e}_y 1/\varepsilon \partial \underline{F}_{TE}/\partial x.$$

Now the electric field has only a $y$-component, that satisfies

$$\Delta \underline{E}_{TE} + \beta_{0,1}^2\,\underline{E}_{TE} = \frac{\partial^2 \underline{E}_{TE}}{\partial x^2} + \frac{\partial^2 \underline{E}_{TE}}{\partial z^2} + \beta_{0,1}^2\,\underline{E}_{TE} = 0\,.$$

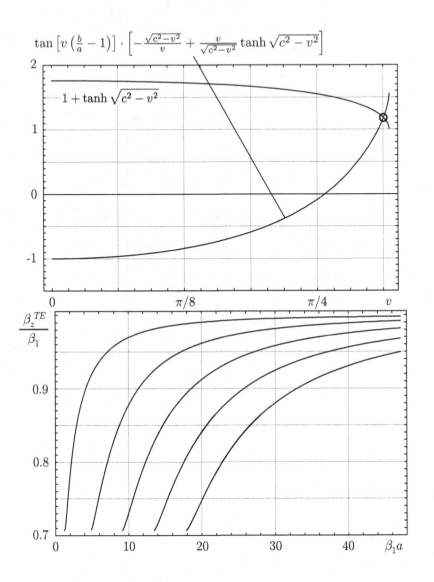

Fig. 6.13–4: Determination of the eigenvalues at $\omega = 3 \cdot 10^{10}$ [s$^{-1}$] and phase constant as a function of the frequency for TE-modes with $\varepsilon_1/\varepsilon_0 = 2$ and $b/a = 2$

$$v^2 = a^2(\beta_1^2 - (\beta_z^{TE})^2); \quad c^2 = a^2(\beta_1^2 - \beta_0^2)$$

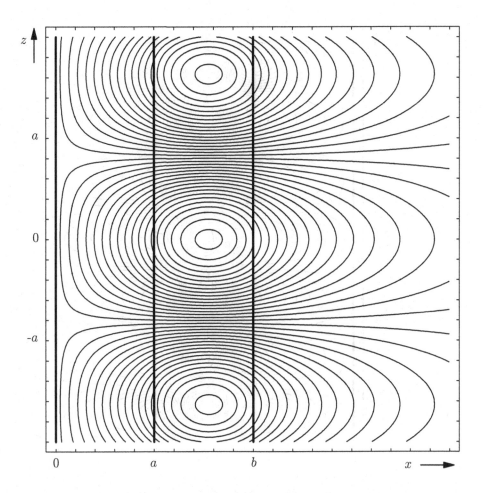

Fig. 6.13–5: Magnetic lines of force of the first TE-Mode for
$$\varepsilon_1/\varepsilon_0 = 2,\ b/a = 2,\ a = 1[\text{cm}],\ \text{and}\ \omega = 5 \cdot 10^{10}\ [\text{s}^{-1}]$$

Thus again the approach for the field is $\underline{E}_{TE} = \underline{E}_0\,\underline{X}_{1,2}(x)\exp(\mp j\beta_z^{TE}z)$ where the functions $\underline{X}_{1,2}(x)$ and the constants have to be redefined.

$$\underline{E}_{TE}(x,z) = \underline{E}_0 \left\{ \begin{array}{l} \underline{C}_1\,\sinh(\alpha x)/\sinh(\alpha a) \\[2mm] \underline{C}_2\,\sin(\beta_x(x-a))+ \\ +\underline{C}_3\,\cos(\beta_x(x-a)) \\[2mm] \underline{C}_4\,\exp(-\alpha(x-b)) \end{array} \right\} \exp(\mp j\beta_z^{TE}z)\,; \quad \begin{array}{l} 0 \le x \le a \\[2mm] a \le x \le b \\[4mm] x \ge b \end{array}$$

$$\alpha^2 = \beta_z^{TE^2} - \beta_0^2 > 0\,; \qquad \beta_x^2 = \beta_1^2 - \beta_z^{TE^2} > 0$$

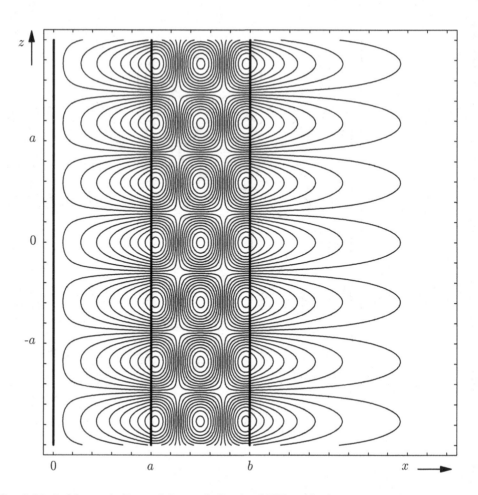

Fig. 6.13–6: Magnetic lines of force of the third TE-mode for
$\varepsilon_1/\varepsilon_0 = 3$, $b/a = 2$, $a = 1[\text{cm}]$, and $\omega = 1.5 \cdot 10^{11}$ $[\text{s}^{-1}]$

$$\vec{\underline{H}}_{TE}(x,z) = -\frac{1}{j\omega\mu}\,\text{rot}\,(\vec{e}_y\,\underline{E}_{TE}) = \frac{1}{j\omega\mu}\left[\vec{e}_x\,\frac{\partial \underline{E}_{TE}}{\partial z} - \vec{e}_z\,\frac{\partial \underline{E}_{TE}}{\partial x}\right]$$

The boundary conditions are

$$\underline{E}_{TE}\big|_{\substack{0<x<a\\x\to a}} = \underline{E}_{TE}\big|_{\substack{a<x<b\\x\to a}} ; \qquad \vec{e}_z\,\vec{\underline{H}}_{TE}\bigg|_{\substack{0<x<a\\x\to a}} = \vec{e}_z\,\vec{\underline{H}}_{TE}\bigg|_{\substack{a<x<b\\x\to a}}$$

$$\underline{E}_{TE}\big|_{\substack{a<x<b\\x\to b}} = \underline{E}_{TE}\big|_{\substack{x>b\\x\to b}} ; \qquad \vec{e}_z\,\vec{\underline{H}}_{TE}\bigg|_{\substack{a<x<b\\x\to b}} = \vec{e}_z\,\vec{\underline{H}}_{TE}\bigg|_{\substack{x>b\\x\to b}}$$

and the evaluation leads to

$$\underline{C}_1 = \underline{C}_3; \qquad \underline{C}_2\sin(\beta_x(b-a)) + \underline{C}_3\cos(\beta_x(b-a)) = \underline{C}_4$$

$$\underline{C}_1 \alpha \coth(\alpha a) = \beta_x \underline{C}_2$$

$$\beta_x \left[ \underline{C}_2 \cos(\beta_x(b-a)) - \underline{C}_3 \sin(\beta_x(b-a)) \right] = -\alpha \underline{C}_4 .$$

Finally this yields the eigenvalue equation

$$\frac{1 + \tanh(\alpha a)}{1 - (\beta_x/\alpha)^2 \tanh(\alpha a)} = -\alpha/\beta_x \tan(\beta_x(b-a))$$

with

$$(\alpha a)^2 + (\beta_x a)^2 = (\beta_1 a)^2 - (\beta_0 a)^2 ; \qquad \beta_z^{TE} = \sqrt{\beta_1^2 - \beta_x^2} = \sqrt{\alpha^2 + \beta_0^2}$$

for the determination of the finite number of solutions for $\beta_x, \beta_z^{TE}$, and $\alpha$.

With $v^2 = a^2(\beta_1^2 - (\beta_z^{TE})^2)$ and $c^2 = a^2(\beta_1^2 - \beta_0^2)$ one gets

$$1 + \tanh\sqrt{c^2 - v^2} = \tan\left[ v\left( \frac{b}{a} - 1 \right) \right] \cdot \left[ -\frac{\sqrt{c^2 - v^2}}{v} + \frac{v}{\sqrt{c^2 - v^2}} \tanh\sqrt{c^2 - v^2} \right] .$$

## 6.14 Layered Dielectric Slab Waveguide

A dielectric slab waveguide consists of a core in $|x| < a$ with permittivity $\varepsilon_2$ and a cladding in $a < |x| < b$ with permittivity $\varepsilon_1 < \varepsilon_2$. The surrounding space has the permittivity $\varepsilon_0$ and the permeability $\mu$ is constant. In the waveguide both transverse magnetic (TM-)waves and transverse electric (TE-)waves can propagate in $z$-direction, while there is no dependence on the $y$-coordinate.

Find the eigenvalue equations for the guided modes whose magnetic (TM) or electric (TE) field is either an even or odd function of the $x$-coordinate.

The waves may be calculated by means of the vector potentials

$$\vec{\underline{A}}_{TM} = \vec{e}_z \underline{A}_{TM}(x, z) \qquad \text{and} \qquad \vec{\underline{F}}_{TE} = \vec{e}_z \underline{F}_{TE}(x, z).$$

As in the present case the electric field (TE-waves) or the magnetic field (TM-waves) has only a $y$-component, it is useful to solve the field equations directly.

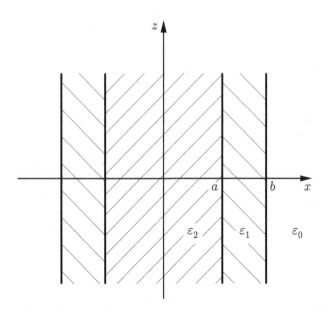

## TM-Waves

The wave propagation with the unknown phase constant $\beta_z^{TM}$ is described by

$$\vec{\underline{H}}_{TM} = \vec{e}_y \underline{H}_{TM}(x,z) ; \qquad \Delta \underline{H}_{TM} + \beta_{0,1,2}^2 \underline{H}_{TM} = 0$$

$$\frac{\partial^2 \underline{H}_{TM}}{\partial x^2} + \frac{\partial^2 \underline{H}_{TM}}{\partial z^2} + \beta_{0,1,2}^2 \underline{H}_{TM} = 0$$

$$\vec{\underline{H}}_{TM}(x,z) = \underline{H}_0 \underline{X}_{0,1,2}(x) \exp(\mp j \beta_z^{TM} z)$$

$$\frac{d^2 \underline{X}_{0,1,2}}{dx^2} + (\beta_{0,1,2}^2 - \beta_z^{TM^2}) \underline{X}_{0,1,2} = 0 ; \qquad \beta_{0,1,2}^2 = \omega^2 \mu \varepsilon_{0,1,2}$$

$$\vec{\underline{E}}_{TM} = \frac{1}{j\omega\varepsilon} \operatorname{rot} \vec{\underline{H}}_{TM} = \frac{1}{j\omega\varepsilon} \left[ -\vec{e}_x \frac{\partial \underline{H}_{TM}}{\partial z} + \vec{e}_z \frac{\partial \underline{H}_{TM}}{\partial x} \right] .$$

In case of even modes the solution functions satisfy $\underline{X}(x) = \underline{X}(-x)$ and the solutio
reads as follows

$$\underline{X}_0(x) = \underline{C}_1 \exp(-\alpha(|x| - b)) ; \qquad\qquad |x| \geq b$$

$$\underline{X}_1(x) = \underline{C}_2 \cos(\beta_{x1}(|x| - a)) + \underline{C}_3 \sin(\beta_{x1}(|x| - a)) ; \qquad a \leq |x| \leq b$$

$$\underline{X}_2(x) = \underline{C}_4 \cos(\beta_{x2} x)/\cos(\beta_{x2} a) ; \qquad\qquad |x| \leq a$$

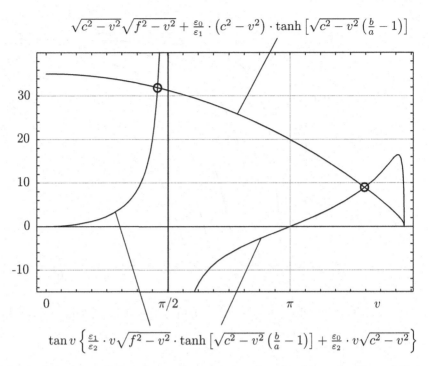

$$\sqrt{c^2 - v^2}\sqrt{f^2 - v^2} + \tfrac{\varepsilon_0}{\varepsilon_1} \cdot \left(c^2 - v^2\right) \cdot \tanh\left[\sqrt{c^2 - v^2}\left(\tfrac{b}{a} - 1\right)\right]$$

$$\tan v \left\{ \tfrac{\varepsilon_1}{\varepsilon_2} \cdot v\sqrt{f^2 - v^2} \cdot \tanh\left[\sqrt{c^2 - v^2}\left(\tfrac{b}{a} - 1\right)\right] + \tfrac{\varepsilon_0}{\varepsilon_2} \cdot v\sqrt{c^2 - v^2} \right\}$$

Fig. 6.14–1: Solutions of the eigenvalue equation for the determination of the phase
constants $\beta_z^{TM}$ of the even TM-modes with
$c = 4.622$, $\varepsilon_2/\varepsilon_0 = 5$, $\varepsilon_1/\varepsilon_0 = 2$, $b/a = 1.5$, and $a = 0.1$ [cm]

with $\beta_z^{TM} = \sqrt{\beta_1^2 - \beta_{x1}^2} = \sqrt{\beta_2^2 - \beta_{x2}^2} = \sqrt{\alpha^2 + \beta_0^2}$.

The other case of odd modes requires $\underline{X}(x) = -\underline{X}(-x)$ and leads to

$$\underline{X}_0(x) = \text{sign}(x)\,\underline{C}_1 \exp(-\alpha(|x| - b)) ; \qquad\qquad |x| \geq b$$

$$\underline{X}_1(x) = \text{sign}(x)\left(\underline{C}_2 \cos(\beta_{x1}(|x| - a)) + \underline{C}_3 \sin(\beta_{x1}(|x| - a))\right); \qquad a \leq |x| \leq b$$

$$\underline{X}_2(x) = \underline{C}_4 \sin(\beta_{x2}x)/\sin(\beta_{x2}a) ; \qquad\qquad |x| \leq a .$$

Here the constants $\underline{C}_{1,2,3,4}$ are redefined in contrast to the first approach for even
modes.

The tangential components of the electric and magnetic field are continuous in $x = a$
and $x = b$. This requires

$$\underline{H}_{TM}\big|_{\substack{x \geq b \\ x \to b}} = \underline{H}_{TM}\big|_{\substack{a \leq x \leq b \\ x \to b}} ; \qquad \underline{H}_{TM}\big|_{\substack{x < a \\ x \to a}} = \underline{H}_{TM}\big|_{\substack{a \leq x \leq b \\ x \to a}}$$

$$\frac{1}{\varepsilon_0}\frac{\partial \underline{H}_{TM}}{\partial x}\bigg|_{\substack{x > b \\ x \to b}} = \frac{1}{\varepsilon_1}\frac{\partial \underline{H}_{TM}}{\partial x}\bigg|_{\substack{a \leq x \leq b \\ x \to b}} ; \qquad \frac{1}{\varepsilon_2}\frac{\partial \underline{H}_{TM}}{\partial x}\bigg|_{\substack{x < a \\ x \to a}} = \frac{1}{\varepsilon_1}\frac{\partial \underline{H}_{TM}}{\partial x}\bigg|_{\substack{a \leq x \leq b \\ x \to a}}$$

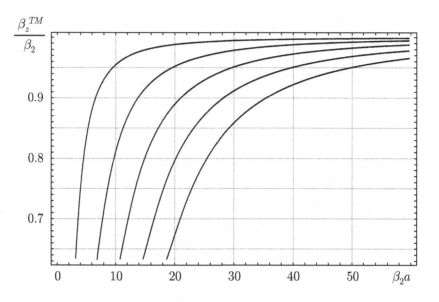

Fig. 6.14–2: Phase constants of the first five odd TM-modes as a function of
$$\beta_2 a = \omega \sqrt{\mu \varepsilon_2} a \text{ with } \varepsilon_1/\varepsilon_0 = 2, \ \varepsilon_2/\varepsilon_0 = 5, \ b/a = 1.5, \text{ and } a = 0.1 \text{ [cm]}$$

For even modes it is

$$\underline{C}_1 = \underline{C}_2 \cos(\beta_{x1}(b-a)) + \underline{C}_3 \sin(\beta_{x1}(b-a)); \qquad \underline{C}_2 = \underline{C}_4$$

$$-\alpha/\varepsilon_0 \, \underline{C}_1 = \beta_{x1}/\varepsilon_1 \left[ -\underline{C}_2 \sin(\beta_{x1}(b-a)) + \underline{C}_3 \cos(\beta_{x1}(b-a)) \right]$$

$$\beta_{x1}/\varepsilon_1 \, \underline{C}_3 = -\beta_{x2}/\varepsilon_2 \, \underline{C}_4 \tan(\beta_{x2}a).$$

Finally this leads to the eigenvalue equation

$$\tan(\beta_{x1}(b-a)) = \frac{\alpha/\beta_{x1} - \beta_{x2}/\beta_{x1} \, \varepsilon_0/\varepsilon_2 \, \tan(\beta_{x2}a)}{\varepsilon_0/\varepsilon_1 + \alpha/\beta_{x1} \, \beta_{x2}/\beta_{x1} \, \varepsilon_1/\varepsilon_2 \, \tan(\beta_{x2}a)}$$

with a finite number of solutions for $\beta_{x1}, \beta_{x2}$, and

$$\beta_z^{TM} = \sqrt{\alpha^2 + \beta_0^2} = \sqrt{\beta_1^2 - \beta_{x1}^2} = \sqrt{\beta_2^2 - \beta_{x2}^2}.$$

As the eigenvalue equation has to be solved numerically, it is convenient to express it in
terms of the variable $v = a\sqrt{\beta_2^2 - (\beta_z^{TM})^2}$ with $f^2 = a^2(\beta_2^2 - \beta_0^2)$ and $c^2 = a^2(\beta_2^2 - \beta_1^2)$

$$v \tan v \left\{ \frac{\varepsilon_1}{\varepsilon_2} \sqrt{f^2 - v^2} \tanh \left[ \sqrt{c^2 - v^2} \left( \frac{b}{a} - 1 \right) \right] + \frac{\varepsilon_0}{\varepsilon_2} \sqrt{c^2 - v^2} \right\} =$$

$$= \sqrt{c^2 - v^2} \sqrt{f^2 - v^2} + \frac{\varepsilon_0}{\varepsilon_1}(c^2 - v^2) \tanh \left[ \sqrt{c^2 - v^2} \left( \frac{b}{a} - 1 \right) \right]$$

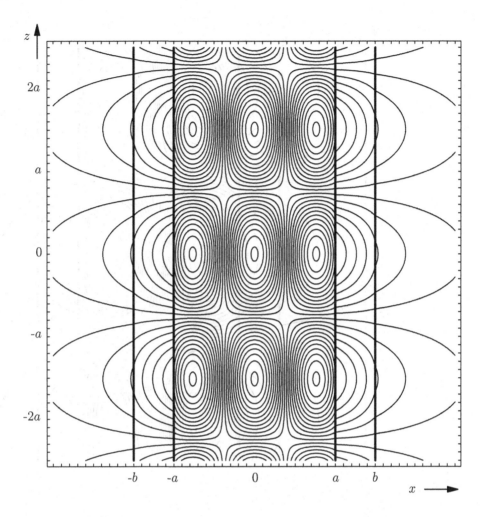

Fig. 6.14–3: Electric lines of force of the second even TM-Mode with
$$\varepsilon_1/\varepsilon_0 = 2, \ \varepsilon_2/\varepsilon_0 = 10, \ b/a = 1.5, \ a = 0.1 \,[\text{cm}], \text{ and } \omega = 4.37 \cdot 10^{11}[s^{-1}]$$

After a similar calculation we obtain the eigenvalue equation for the odd TM-modes. The equations are

$$\underline{C}_1 = \underline{C}_2 \cos(\beta_{x1}(b-a)) + \underline{C}_3 \sin(\beta_{x1}(b-a)); \quad \underline{C}_2 = \underline{C}_4$$

$$-\alpha/\varepsilon_0 \,\underline{C}_1 = \beta_{x1}/\varepsilon_1 \left[ -\underline{C}_2 \sin(\beta_{x1}(b-a)) + \underline{C}_3 \cos(\beta_{x1}(b-a)) \right]$$

$$\beta_{x1}/\varepsilon_1 \,\underline{C}_3 = \beta_{x2}/\varepsilon_2 \,\underline{C}_4 \cot(\beta_{x2}a)$$

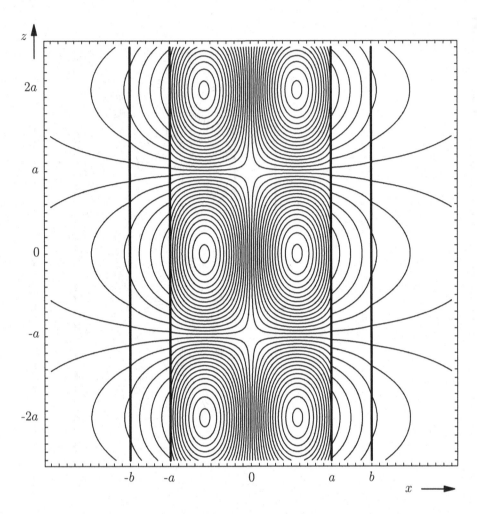

Fig. 6.14–4: Electric lines of first of the first odd TM-mode with
$\varepsilon_1/\varepsilon_0 = 2$, $\varepsilon_2/\varepsilon_0 = 10$, $b/a = 1.5$, $a = 0.1$ [cm], and $\omega = 2.99 \cdot 10^{11}$ [s$^{-1}$

and the eigenvalue equation becomes

$$\tan(\beta_{x1}(b - a)) = \frac{\alpha/\beta_{x1} + \varepsilon_0/\varepsilon_2\,\beta_{x2}/\beta_{x1}\,\cot(\beta_{x2}a)}{\varepsilon_0/\varepsilon_1 - \alpha/\beta_{x1}\,\varepsilon_1/\varepsilon_2\,\beta_{x2}/\beta_{x1}\,\cot(\beta_{x2}a)}.$$

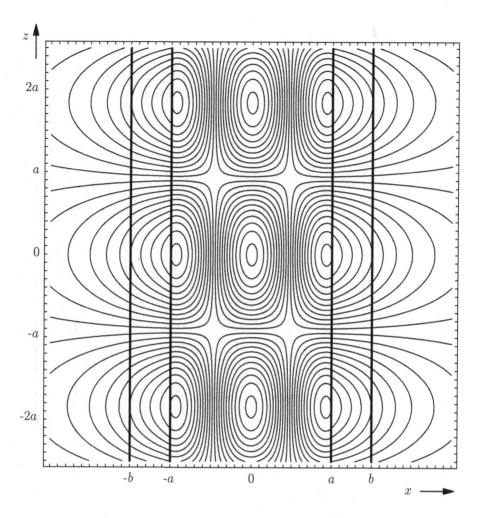

Fig. 6.14–5: Magnetic lines of force of the second even TE-mode with
$\varepsilon_1/\varepsilon_0 = 2$, $\varepsilon_2/\varepsilon_0 = 10$, $b/a = 1.5$, $a = 0.1$ [cm], and $\omega = 3.58 \cdot 10^{11}$ [s$^{-1}$]

**TE-Waves**

The analog approach for the electric field of the TE-modes is

$$\vec{\underline{E}}_{TE}(x, z) = \underline{E}_0\underline{X}_{0,1,2}(x) \exp(\mp j\beta_z^{TE} z)$$

The functions $\underline{X}_{0,1,2}$ remain unchanged but with new constants $\underline{C}_{1,2,3,4}$.

After evaluating the boundary conditions we finally get the eigenvalue equations

$$\tan(\beta_{x1}(b - a)) = \frac{\alpha/\beta_{x1} - \beta_{x2}/\beta_{x1}\tan(\beta_{x2}a)}{1 + \alpha/\beta_{x1}\beta_{x2}/\beta_{x1}\tan(\beta_{x2}a)} \ ; \qquad \text{(even TE-modes)}$$

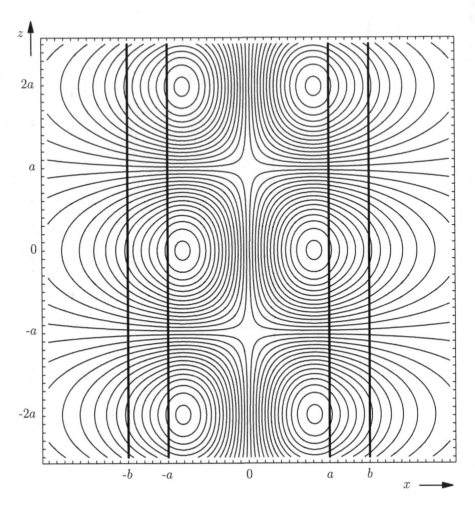

Fig. 6.14–6: Magnetic lines of force of the first odd TE-Mode with
$\varepsilon_1/\varepsilon_0 = 2$, $\varepsilon_2/\varepsilon_0 = 5$, $b/a = 1.5$, $a = 0.1$ [cm], and $\omega = 3.35 \cdot 10^{11}$ [s$^{-1}$]

$$\tan(\beta_{x1}(b-a)) = \frac{\alpha/\beta_{x1} + \beta_{x2}/\beta_{x1}\cot(\beta_{x2}a)}{1 - \alpha/\beta_{x1}\beta_{x2}/\beta_{x1}\cot(\beta_{x2}a)} \; ; \qquad \text{(odd TE-modes)}.$$

## 6.15 Diffraction by a Dielectric Cylinder

A dielectric cylinder has the radius $a$ and is of infinite length in $z$-direction. Its permeability and permittivity are $\mu$ and $\varepsilon$, whereas the surrounding medium is the free space with material parameters $\mu_0$ and $\varepsilon_0$.

1. Calculate the electromagnetic field excited by a plane wave, when the phase vector $\vec{\beta}_0 = -\vec{e}_x \beta_0$; $\beta_0 = \omega \sqrt{\mu_0 \varepsilon_0}$ is directed perpendicular to the axis of the cylinder. In the first case a) the wave is $z$–polarized and in the second case b) $y$–polarized.

2. Find the electromagnetic field excited by a line source $i(t) = \mathrm{Re}\,\{\underline{i}_0 \exp(j\omega t)\}$ with distance $c > a$ parallel to the axis of the cylinder.

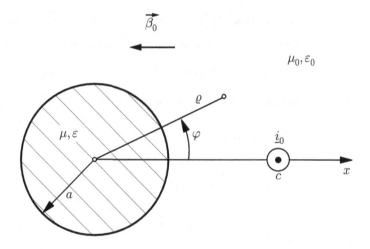

**1.a) The Exciting Field is $z$-Polarized**

$$\vec{\underline{E}}_0 = \vec{e}_z\, E_0\, \exp(j\beta_0 x) = \vec{e}_z\, E_0\, \exp(j\beta_0 \varrho \cos\varphi)\,.$$

The series expansion in terms of Bessel functions is

$$\exp(j\beta_0 \varrho \cos\varphi) = J_0\,(\beta_0 \varrho) + 2 \sum_{n=1}^{\infty} \exp(jn\pi/2)\, J_n\,(\beta_0 \varrho) \cos(n\varphi)$$

$$= \sum_{n=0}^{\infty} (2 - \delta_{0n}) \exp(jn\pi/2)\, J_n\,(\beta_0 \varrho) \cos(n\varphi)\,.$$

With it the exciting field is of the form

$$\vec{\underline{E}}_0 = \vec{e}_z \, E_0 \sum_{n=0}^{\infty} (2 - \delta_{0n}) \, \exp(jn\pi/2) \, J_n \,(\beta_0 \varrho) \cos(n\varphi) \, .$$

Here $\delta_{mn}$ denotes the Kronecker delta defined by

$$\delta_{mn} = \begin{cases} 0; & m \neq n \\ 1; & m = n \, . \end{cases}$$

The diffracted wave is described by Hankel functions $H_n^{(2)}(\beta_0 \varrho)$ and inside the cylinder the field is described by Bessel functions $J_n(\beta \varrho)$ with $\beta = \omega \sqrt{\mu \varepsilon}$. Hence the resulting electric field is

$$\vec{\underline{E}} = \vec{e}_z \, E_0 \sum_{n=0}^{\infty} (2-\delta_{0n}) \, \exp(jn\pi/2) \begin{Bmatrix} J_n \,(\beta_0 \varrho) + \underline{a}_n \, H_n^{(2)} \,(\beta_0 \varrho) \\ \underline{b}_n \, J_n(\beta \varrho) \end{Bmatrix} \cos(n\varphi) \, ; \quad \begin{matrix} \varrho \geq a \\ \varrho \leq a \end{matrix}$$

where the constants $\underline{a}_n, \underline{b}_n$ follow from the boundary conditions in $\varrho = a$.

The continuity of the electric field at $\varrho = a$ leads to

$$J_n(\beta_0 a) + \underline{a}_n \, H_n^{(2)}(\beta_0 a) = \underline{b}_n \, J_n(\beta a) \, ; \qquad H_n^{(2)}(\beta_0 \varrho) = J_n(\beta_0 \varrho) - j N_n(\beta_0 \varrho)$$

and the continuity of the tangential component $\underline{H}_\varphi$ of the magnetic field yields

$$\mathrm{rot} \, [\vec{e}_z \underline{E}(\varrho, \varphi)] = \mathrm{grad} \, \underline{E} \times \vec{e}_z = \vec{e}_\varrho \frac{1}{\varrho} \frac{\partial \underline{E}}{\partial \varphi} - \vec{e}_\varphi \frac{\partial \underline{E}}{\partial \varrho} = -j\omega\mu \, \vec{\underline{H}}$$

$$\underline{H}_\varrho = -\frac{1}{j\omega\mu} \frac{1}{\varrho} \frac{\partial \underline{E}}{\partial \varphi} \, ; \qquad \underline{H}_\varphi = \frac{1}{j\omega\mu} \frac{\partial \underline{E}}{\partial \varrho}$$

$$\frac{1}{\mu} \frac{\partial \underline{E}}{\partial \varrho} \bigg|_{\substack{\varrho < a \\ \varrho \to a}} = \frac{1}{\mu_0} \frac{\partial \underline{E}}{\partial \varrho} \bigg|_{\substack{\varrho > a \\ \varrho \to a}}$$

$$\Rightarrow \quad \frac{\beta_0}{\mu_0} \left[ J_n'(\beta_0 a) + \underline{a}_n H_n^{(2)\prime}(\beta_0 a) \right] = \frac{\beta}{\mu} \left[ \underline{b}_n J_n'(\beta a) \right] \, .$$

Thus, there are two equations for the determination of the constants $\underline{a}_n$ and $\underline{b}_n$.

$$\underline{a}_n = \underline{M}_n^{-1} \left[ \frac{Z_0}{Z} J_n(\beta_0 a) J_n'(\beta a) - J_n'(\beta_0 a) J_n(\beta a) \right]$$

$$\underline{b}_n = \underline{M}_n^{-1} \left[ H_n^{(2)\prime}(\beta_0 a) J_n(\beta_0 a) - H_n^{(2)}(\beta_0 a) J_n'(\beta_0 a) \right]$$

$$\text{with} \quad \underline{M}_n = H_n^{(2)\prime}(\beta_0 a) J_n(\beta a) - \frac{Z_0}{Z} H_n^{(2)}(\beta_0 a) J_n'(\beta a)$$

$$Z = \sqrt{\frac{\mu}{\varepsilon}} \, ; \qquad Z_0 = \sqrt{\frac{\mu_0}{\varepsilon_0}}$$

In the special case of a perfect conducting cylinder the field in $\varrho < a$ vanishes.

$$E_z\big|_{\varrho=a} = 0 \; ; \qquad H_\varrho\big|_{\varrho=a} = 0$$

$$J_n(\beta_0 a) + \underline{a}_n \, H_n^{(2)}(\beta_0 a) = 0 \; ; \qquad \underline{a}_n = -\frac{J_n(\beta_0 a)}{H_n^{(2)}(\beta_0 a)}$$

Now the surface of the cylinder $\varrho = a$ carries a current sheet $\vec{\underline{K}}_E = \vec{e}_z \underline{K}_E$. It follows

$$\underline{H}_\varphi\big|_{\substack{\varrho>a \\ \varrho \to a}} - \underline{H}_\varphi\big|_{\substack{\varrho<a \\ \varrho \to a}} = \underline{K}_E$$

$$\underline{K}_E = \underline{H}_\varphi\big|_{\substack{\varrho>a \\ \varrho\to a}} = \frac{1}{j\omega\mu_0} \frac{\partial \underline{E}}{\partial \varrho}\bigg|_{\substack{\varrho>a \\ \varrho\to a}} =$$

$$= -\frac{j\beta_0}{\omega\mu_0} E_0 \sum_{n=0}^{\infty} (2-\delta_{0n}) \exp(jn\pi/2) \left[ J_n'(\beta_0 a) - \frac{J_n(\beta_0 a)}{H_n^{(2)}(\beta_0 a)} H_n^{(2)'}(\beta_0 a) \right] \cos(n\varphi)$$

Finally with

$$J_n'(z) = -J_{n+1}(z) + \frac{n}{z} J_n(z) \; ; \qquad H_n^{(2)'}(z) = -H_{n+1}^{(2)}(z) + \frac{n}{z} H_n^{(2)}(z)$$

$$H_n^{(2)}(z) = J_n(z) - jN_n(z) \; ;$$

$$J_{n+1}(z)N_n(z) - J_n(z)\,N_{n+1}(z) = \frac{2}{\pi z} \quad \text{(Wronskian determinant)}$$

$$\Rightarrow \quad \underline{K}_E = \frac{2E_0}{\pi\beta_0 a Z_0} \sum_{n=0}^{\infty} \frac{(2-\delta_{0n})\exp(jn\pi/2)}{H_n^{(2)}(\beta_0 a)} \cos(n\varphi) .$$

## 1.b) Orthogonal Polarization Compared to 1.a)

Now the exciting magnetic field is $z$-directed

$$\vec{\underline{H}} = \vec{e}_z \, H_0 \, \exp(j\beta_0 \varrho \cos\varphi)$$

and accordingly the resulting magnetic field is

$$\vec{\underline{H}} = \vec{e}_z \, H_0 \sum_{n=0}^{\infty} (2-\delta_{0n}) \exp(jn\pi/2) \left\{ \begin{array}{l} J_n(\beta_0 \varrho) + \underline{c}_n H_n^{(2)}(\beta_0 \varrho) \\ \underline{d}_n \, J_n(\beta\varrho) \end{array} \right\} \cos(n\varphi) \; ; \quad \begin{array}{l} \varrho \geq a \\ \varrho \leq a \end{array} .$$

Again, the constants $\underline{c}_n$ and $\underline{d}_n$ follow from the boundary conditions in $\varrho = a$.

The continuity of the magnetic field leads to

$$J_n(\beta_0 a) + \underline{c}_n \, H_n^{(2)}(\beta_0 a) = \underline{d}_n \, J_n(\beta a)$$

and the continuity of the tangential component $\underline{E}_\varphi$ of

$$\vec{\underline{E}} = \frac{1}{j\omega\varepsilon}\mathrm{rot}\,[\vec{e}_z\underline{H}(\varrho,\varphi)] = \frac{1}{j\omega\varepsilon}\left(\vec{e}_\varrho\frac{1}{\varrho}\frac{\partial\underline{H}}{\partial\varphi} - \vec{e}_\varphi\frac{\partial\underline{H}}{\partial\varrho}\right)$$

$$\underline{E}_\varrho = \frac{1}{j\omega\varepsilon}\frac{1}{\varrho}\frac{\partial\underline{H}}{\partial\varphi}\;;\qquad \underline{E}_\varphi = -\frac{1}{j\omega\varepsilon}\frac{\partial\underline{H}}{\partial\varrho}$$

yields

$$\frac{1}{\varepsilon}\frac{\partial\underline{H}}{\partial\varrho}\bigg|_{\substack{\varrho<a\\\varrho\to a}} = \frac{1}{\varepsilon_0}\frac{\partial\underline{H}}{\partial\varrho}\bigg|_{\substack{\varrho>a\\\varrho\to a}}$$

$$J_n'(\beta_0 a) + \underline{c}_n H_n^{(2)\prime}(\beta_0 a) = \frac{\beta}{\beta_0}\frac{\varepsilon_0}{\varepsilon}\underline{d}_n J_n'(\beta a) = \frac{Z}{Z_0}\underline{d}_n J_n'(\beta a)\,.$$

Hence the constants are

$$\underline{M}_n = H_n^{(2)\prime}(\beta_0 a)J_n(\beta a) - \frac{Z}{Z_0}H_n^{(2)}(\beta_0 a)J_n'(\beta a)$$

$$\underline{c}_n = \underline{M}_n^{-1}\left[\frac{Z}{Z_0}J_n(\beta_0 a)J_n'(\beta a) - J_n(\beta a)J_n'(\beta_0 a)\right]$$

$$\underline{d}_n = \underline{M}_n^{-1}\left[H_n^{(2)\prime}(\beta_0 a)J_n(\beta_0 a) - H_n^{(2)}(\beta_0 a)J_n'(\beta_0 a)\right]\,.$$

In the limit of a perfect conducting cylinder it is

$$\underline{E}_\varphi\big|_{\substack{\varrho>a\\\varrho\to a}} = 0\,;\qquad \underline{c}_n = -\frac{J_n'(\beta_0 a)}{H_n^{(2)\prime}(\beta_0 a)}$$

and the current sheet on the surface of the cylinder is given by

$$\vec{\underline{K}}_H = \vec{e}_\varphi\underline{K}_H\,;\qquad \underline{H}\big|_{\substack{\varrho<a\\\varrho\to a}} - \underline{H}\big|_{\substack{\varrho>a\\\varrho\to a}} = \underline{K}_H$$

$$\underline{K}_H = -\underline{H}\big|_{\varrho=a}$$

$$= H_0\sum_{n=0}^{\infty}(2-\delta_{0n})\exp(jn\pi/2)\cdot$$

$$\cdot\left[\frac{J_n'(\beta_0 a)}{H_n^{(2)\prime}(\beta_0 a)}H_n^{(2)}(\beta_0 a) - J_n(\beta_0 a)\right]\cos(n\varphi)$$

$$\underline{K}_H = j\frac{2H_0}{\pi\beta_0 a}\sum_{n=0}^{\infty}\frac{(2-\delta_{0n})\exp(jn\pi/2)}{H_n^{(2)\prime}(\beta_0 a)}\cos(n\varphi)\,.$$

**2.) Field Excitation by a Line Source $i(t)$ at Position $(\varrho = c > a, \varphi = \varphi_0)$**

The vector potential of a line source in the homogeneous space with $\mu_0, \varepsilon_0$ is:

$$\vec{\underline{A}}_0 = \vec{e}_z\underline{A}_0(\varrho,\varphi)\,;\qquad \Delta\underline{A}_0 + \beta_0^2\underline{A}_0 = 0\,;\qquad \varrho\neq c$$

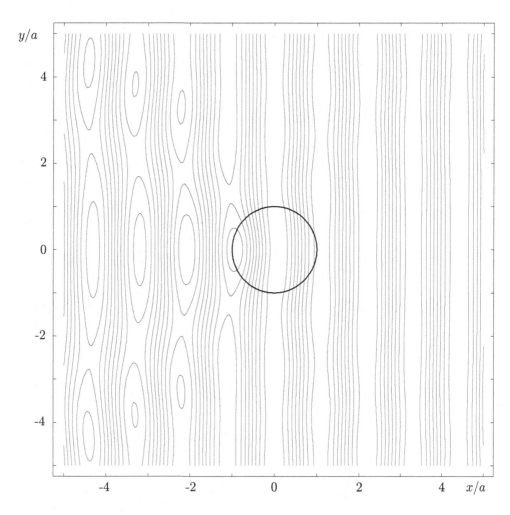

Fig. 6.15–1: Diffraction of a plane, $z$-polarized wave: Magnetic lines of force for
$\lambda_0/\lambda = 1.1$ and $\lambda/a = 2$ $(\lambda_0 = 2\,\pi/\beta_0;\ \lambda = 2\pi/\beta)$

$$\underline{A}_0(\varrho,\varphi) = \sum_{n=0}^{\infty} \underline{a}_n \left\{ \begin{array}{l} J_n(\beta_0 c)\, H_n^{(2)}(\beta_0 \varrho) \\ H_n^{(2)}(\beta_0 c)\, J_n(\beta_0 \varrho) \end{array} \right\} \cos(n(\varphi - \varphi_0));\quad \begin{array}{l} \varrho \geq c \\ \varrho \leq c \end{array}\ .$$

This approach already satisfies the continuity relation of the electric field and the
continuity of the radial component of $\vec{\underline{B}}_0 = \operatorname{rot} \vec{\underline{A}}_0$ at $\varrho = c$.

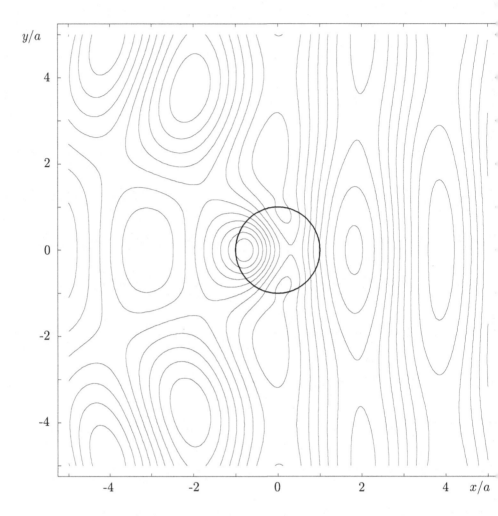

Fig. 6.15–2: See Fig. 6.15–1, but $\lambda_0/\lambda = 2$

For the tangential component of the magnetic field applies

$$\underline{H}_{0\varphi}\Big|_{\substack{\varrho>c\\ \varrho\to c}} - \underline{H}_{0\varphi}\Big|_{\substack{\varrho<c\\ \varrho\to c}} = \underline{K}(\varphi) = \frac{1}{\mu_0}\left[\frac{\partial\underline{A}_0}{\partial\varrho}\Big|_{\substack{\varrho<c\\ \varrho\to c}} - \frac{\partial\underline{A}_0}{\partial\varrho}\Big|_{\substack{\varrho>c\\ \varrho\to c}}\right]$$

$$\underline{K}(\varphi) = \frac{\beta_0}{\mu_0}\sum_{n=0}^{\infty}\underline{a}_n\left[H_n^{(2)}(\beta_0 c)J_n'(\beta_0 c) - J_n(\beta_0 c)H_n^{(2)\prime}(\beta_0 c)\right]\cos(n(\varphi-\varphi_0)).$$

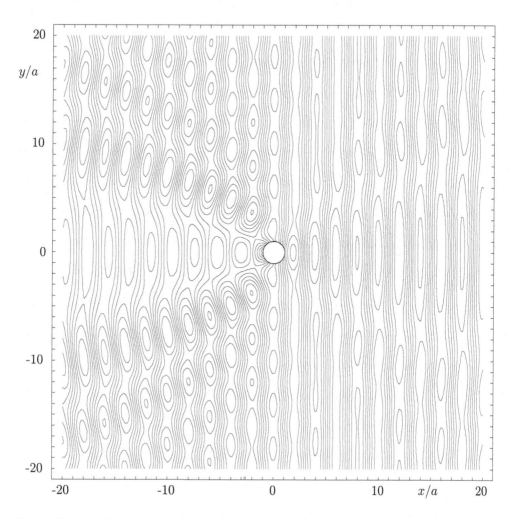

Fig. 6.15–3: See Fig. 6.15–2

With the orthogonality relation for trigonometric functions one gets:

$$\beta_0 \underline{a}_n \left[ (J_n - jN_n) \left( -J_{n+1} + \frac{n}{\beta_0 c} J_n \right) - J_n \left( -H^{(2)}_{n+1} + \frac{n}{\beta_0 c} H_n \right) \right] \pi (1 + \delta_{0n}) =$$

$$= \frac{\mu_0}{c} \int\limits_{0}^{2\pi} \underline{K}(\varphi) \cos(n(\varphi - \varphi_0)) c\, d\varphi = \frac{\mu_0}{c} \underline{i}_0 =$$

$$= \beta_0 \underline{a}_n \pi (1 + \delta_{0n}) j \left[ J_{n+1} N_n - J_n N_{n+1} \right] = 2j \frac{\underline{a}_n (1 + \delta_{0n})}{c} = \frac{\mu_0}{c} \underline{i}_0$$

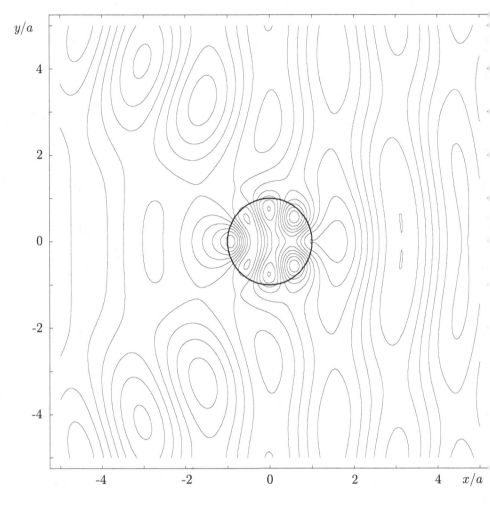

Fig. 6.15–4: See Fig. 6.15–1, but $\lambda_0/\lambda = 3$; $\lambda/a = 1$

and the unknown constants can be calculated

$$\Rightarrow \quad \underline{a}_n = -j\,\frac{\mu_0 \underline{i}_0}{2(1+\delta_{0n})}\,.$$

Thus, the exciting vector potential is

$$\underline{A}_0(\varrho,\varphi) = -j\,\frac{\mu_0 \underline{i}_0}{2} \sum_{n=0}^{\infty} \frac{1}{1+\delta_{0n}} \left\{ \begin{array}{l} J_n(\beta_0 c)\,H_n^{(2)}(\beta_0 \varrho) \\[4pt] H_n^{(2)}(\beta_0 c)\,J_n(\beta_0 \varrho) \end{array} \right\} \cos(n(\varphi-\varphi_0))\,; \qquad \begin{array}{l} \varrho \geq c \\[4pt] \varrho \leq c \end{array}$$

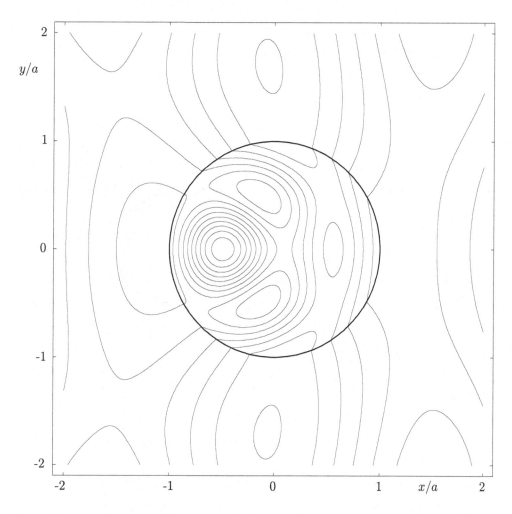

Fig. 6.15–5: Diffraction of a plane, $y$-polarized wave at a dielectric cylinder: Electric
lines of force for $\lambda_0/\lambda = 3$ and $\lambda/a = 1$

An approach for the resulting field in presence of the cylinder is:

$$\vec{\underline{A}} = \vec{e}_z\underline{A}(\varrho,\varphi)\;; \qquad \Delta\underline{A} + \left\{ \begin{array}{c} \beta_0^2 \\ \beta^2 \end{array} \right\} \underline{A} = 0\;; \qquad \begin{array}{c} \varrho > a\,;\, \varrho \neq c \\ \varrho < a \end{array}$$

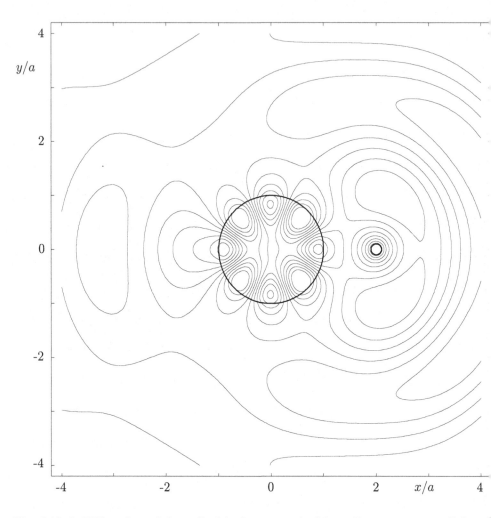

Fig. 6.15–6: Diffraction of the cylindrical wave excited by a line source at a dielectr
cylinder: Magnetic lines of force for $\lambda_0/\lambda = 3$, $\lambda/a = 1$, and $c/a = 2$

$$\underline{A}(\varrho, \varphi) = -j\frac{\mu_0}{2}\underline{i}_0 \sum_{n=0}^{\infty} \frac{1}{1 + \delta_{0n}} \cos(n(\varphi - \varphi_0)) \cdot$$

$$\cdot \left\{ \begin{array}{ll} \underline{b}_n \, J_n(\beta\varrho) & ; \ \varrho < a \\ \underline{c}_n \, H_n^{(2)}(\beta_0\varrho) + \left\{ \begin{array}{ll} J_n(\beta_0 c)H_n^{(2)}(\beta_0\varrho); & \varrho > c \\ H_n^{(2)}(\beta_0 c)J_n(\beta_0\varrho); & \varrho < c \end{array} \right\} & ; \ \varrho > a \end{array} \right\} \cdot$$

The constants follow from the boundary conditions in $\varrho = a$.

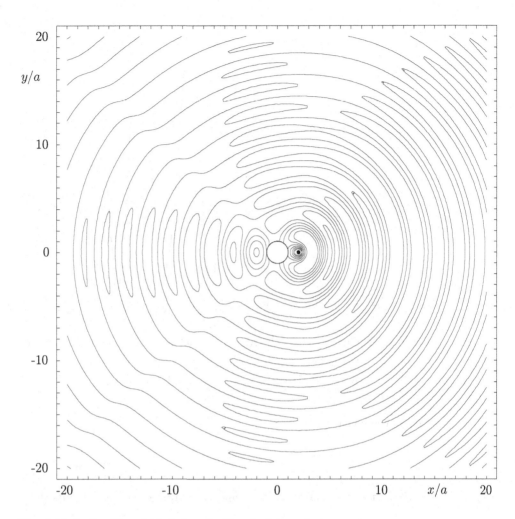

Fig. 6.15–7: See Fig. 6.15–6; but $\lambda_0/\lambda = 2$; $\lambda/a = 2$

At first the continuity of the normal component of the magnetic flux density leads to

$$\underline{b}_n J_n(\beta a) = \underline{c}_n H_n^{(2)}(\beta_0 a) + H_n^{(2)}(\beta_0 c) J_n(\beta_0 a)$$

and a second equation results from the continuity of the tangential component of the magnetic field:

$$\underline{b}_n J_n'(\beta a) = \frac{\beta_0}{\beta} \frac{\mu}{\mu_0} \left[ \underline{c}_n H_n^{(2)\prime}(\beta_0 a) + H_n^{(2)}(\beta_0 c) J_n'(\beta_0 a) \right] .$$

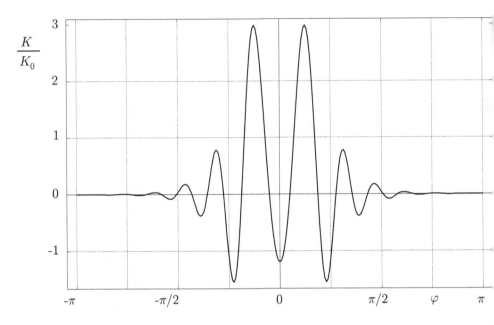

Fig. 6.15–8:  Current sheet $K/K_0$ on the surface of a conducting cylinder exposed to
the field of a line source at $(\varrho = c,\ \varphi_0 = 0)$ with
$K_0 = -|\underline{i}_0|/(\pi a),\ \lambda_0/\lambda = 1.1$, and $\lambda/a = 1/2$

Hence, the constants $\underline{b}_n$ are

$$\underline{b}_n = \frac{H_n^{(2)}(\beta_0 c)}{\underline{M}_n}\left[J_n(\beta_0 a)H_n^{(2)\prime}(\beta_0 a) - J_n'(\beta_0 a)\,H_n^{(2)}(\beta_0 a)\right].$$

Applying the recurrence relations and the Wronskian-determinants they become

$$\underline{b}_n = -j\frac{2}{\pi}\frac{H_n^{(2)}(\beta_0 c)}{\beta_0 a \underline{M}_n}$$

with

$$\underline{M}_n = J_n(\beta a)H_n^{(2)\prime}(\beta_0 a) - \frac{Z_0}{Z}\,J_n'(\beta a)\,H_n^{(2)}(\beta_0 a).$$

For the constants $\underline{c}_n$ it follows

$$\underline{c}_n = -\frac{H_n^{(2)}(\beta_0 c)}{\underline{M}_n}\left[J_n(\beta a)\,J_n'(\beta_0 a) - \frac{Z_0}{Z}\,J_n'(\beta a)\,J_n(\beta_0 a)\right].$$

In the limit $\mu = \mu_0;\quad \varepsilon = \varepsilon_0;\quad \beta = \beta_0;\quad Z = Z_0$ it is

$$\underline{c}_n\big|_{\substack{\mu=\mu_0\\ \varepsilon=\varepsilon_0}} = 0;\qquad \underline{b}_n\big|_{\substack{\mu=\mu_0\\ \varepsilon=\varepsilon_0}} = H_n^{(2)}(\beta_0 c);$$

$$\underline{M}_n\Big|_{\substack{\mu=\mu_0 \\ \varepsilon=\varepsilon_0}} = -j\,\frac{2}{\pi\beta_0 a}$$

and thus, just for the purpose of validation, we again get the exciting field of the line source in free space.

If the cylinder is a perfect conductor, then the tangential component of the electric field vanishes at $\varrho = a$ and this leads to

$$\underline{c}_n H_n^{(2)}(\beta_0 a) + H_n^{(2)}(\beta_0 c)\, J_n(\beta_0 a) = 0$$

$$\Rightarrow \quad \underline{c}_n = -J_n(\beta_0 a)\,\frac{H_n^{(2)}(\beta_0 c)}{H_n^{(2)}(\beta_0 a)}\,.$$

Therewith the current sheet on the surface of the cylinder is

$$\vec{\underline{K}} = \vec{e}_z \underline{K} = \vec{e}_z\,\underline{H}_\varphi\Big|_{\substack{\varrho>a \\ \varrho\to a}} = -\vec{e}_z\,\frac{1}{\mu_0}\,\frac{\partial\underline{A}}{\partial\varrho}\Big|_{\varrho=a}$$

$$\underline{K} = j\frac{i_0}{2}\beta_0\sum_{n=0}^{\infty}\frac{1}{1+\delta_{0n}}\left[-J_n(\beta_0 a)\frac{H_n^{(2)}(\beta_0 c)}{H_n^{(2)}(\beta_0 a)}H_n^{(2)\prime}(\beta_0 a)+\right.$$

$$\left. + H_n^{(2)}(\beta_0 c)J_n'(\beta_0 a)\right]\cdot\cos(n(\varphi-\varphi_0))$$

and with the recurrence relations and the Wronskian-determinants this expression simplifies to

$$\underline{K} = -\frac{i_0}{\pi a}\sum_{n=0}^{\infty}\frac{1}{1+\delta_{0n}}\frac{H_n^{(2)}(\beta_0 c)}{H_n^{(2)}(\beta_0 a)}\cos(n(\varphi-\varphi_0))\,.$$

# Appendix

## Vector Calculus

### Multiple Products of Vectors

$$\vec{A}(\vec{B} \times \vec{C}) = \vec{C}(\vec{A} \times \vec{B}) = \vec{B}(\vec{C} \times \vec{A})$$
$$\vec{A} \times (\vec{B} \times \vec{C}) = \vec{B}(\vec{A}\vec{C}) - \vec{C}(\vec{A}\vec{B})$$
$$(\vec{A} \times \vec{B})(\vec{C} \times \vec{D}) = (\vec{A}\vec{C})(\vec{B}\vec{D}) - (\vec{A}\vec{D})(\vec{B}\vec{C})$$

### Vector Operations in Arbitrary Orthogonal Coordinates $(u_1, u_2, u_3)$

Metric factors: $h_1, h_2, h_3$

$$V = V(u_1, u_2, u_3);$$
$$\vec{A} = \vec{e}_1 A_1(u_1, u_2, u_3) + \vec{e}_2 A_2(u_1, u_2, u_3) + \vec{e}_3 A_3(u_1, u_2, u_3)$$
$$\operatorname{grad} V = \vec{e}_1 \frac{1}{h_1} \frac{\partial V}{\partial u_1} + \vec{e}_2 \frac{1}{h_2} \frac{\partial V}{\partial u_2} + \vec{e}_3 \frac{1}{h_3} \frac{\partial V}{\partial u_3}$$
$$\operatorname{div} \vec{A} = \frac{1}{h_1 h_2 h_3} \left[ \frac{\partial}{\partial u_1}(h_2 h_3 A_1) + \frac{\partial}{\partial u_2}(h_3 h_1 A_2) + \frac{\partial}{\partial u_3}(h_1 h_2 A_3) \right]$$

$$\operatorname{rot} \vec{A} = \begin{vmatrix} \vec{e}_1/(h_2 h_3) & \vec{e}_2/(h_1 h_3) & \vec{e}_3/(h_1 h_2) \\ \partial/\partial u_1 & \partial/\partial u_2 & \partial/\partial u_3 \\ h_1 A_1 & h_2 A_2 & h_3 A_3 \end{vmatrix}$$

$$\operatorname{div} \operatorname{grad} V = \Delta V =$$

$$= \frac{1}{h_1 h_2 h_3} \left[ \frac{\partial}{\partial u_1}\left( \frac{h_2 h_3}{h_1} \frac{\partial V}{\partial u_1} \right) + \frac{\partial}{\partial u_2}\left( \frac{h_1 h_3}{h_2} \frac{\partial V}{\partial u_2} \right) + \frac{\partial}{\partial u_3}\left( \frac{h_1 h_2}{h_3} \frac{\partial V}{\partial u_3} \right) \right]$$

Cartesian coordinates:

$$u_1 = x; \qquad u_2 = y; \qquad u_3 = z; \qquad h_1 = h_2 = h_3 = 1$$

$$\vec{r} = \vec{e}_x x + \vec{e}_y y + \vec{e}_z z; \quad d\vec{r} = \vec{e}_x\, dx + \vec{e}_y\, dy + \vec{e}_z\, dz; \quad dv = dx\, dy\, dz$$

$$\vec{A}(\vec{r}) \;=\; \vec{e}_x\,A_x + \vec{e}_y\,A_y + \vec{e}_z\,A_z\,; \qquad V = V(x,y,z)$$

$$\mathrm{grad}\,V \;=\; \vec{e}_x\,\frac{\partial V}{\partial x} + \vec{e}_y\,\frac{\partial V}{\partial y} + \vec{e}_z\,\frac{\partial V}{\partial z}$$

$$\mathrm{div}\,\vec{A} \;=\; \frac{\partial A_x}{\partial x} + \frac{\partial A_y}{\partial y} + \frac{\partial A_z}{\partial z}$$

$$\mathrm{rot}\,\vec{A} \;=\; \vec{e}_x\left(\frac{\partial A_z}{\partial y} - \frac{\partial A_y}{\partial z}\right) + \vec{e}_y\left(\frac{\partial A_x}{\partial z} - \frac{\partial A_z}{\partial x}\right) + \vec{e}_z\left(\frac{\partial A_y}{\partial x} - \frac{\partial A_x}{\partial y}\right)$$

$$\Delta V \;=\; \frac{\partial^2 V}{\partial x^2} + \frac{\partial^2 V}{\partial y^2} + \frac{\partial^2 V}{\partial z^2}$$

Circular-cylinder coordinates:

$$u_1 = \varrho\,; \quad u_2 = \varphi\,; \quad u_3 = z\,; \quad h_1 = 1\,; \quad h_2 = \varrho\,; \quad h_3 = 1$$

$$\vec{r} = \vec{e}_\varrho(\varphi)\,\varrho + \vec{e}_z z\,; \quad d\vec{r} = \vec{e}_\varrho\,d\varrho + \vec{e}_\varphi\,\varrho\,d\varphi + \vec{e}_z\,dz\,; \quad dv = \varrho\,d\varrho\,d\varphi\,dz$$

$$\vec{A}(\vec{r}) = \vec{e}_\varrho(\varphi)\,A_\varrho + \vec{e}_\varphi(\varphi)\,A_\varphi + \vec{e}_z\,A_z\,; \quad V = V(\varrho,\varphi,z)$$

$$x = \varrho\cos\varphi\,; \quad y = \varrho\sin\varphi\,; \quad z = z$$

$$\mathrm{grad}\,V = \vec{e}_\varrho\,\frac{\partial V}{\partial \varrho} + \frac{\vec{e}_\varphi}{\varrho}\,\frac{\partial V}{\partial \varphi} + \vec{e}_z\,\frac{\partial V}{\partial z}$$

$$\mathrm{div}\,\vec{A} = \frac{1}{\varrho}\,\frac{\partial}{\partial \varrho}(\varrho A_\varrho) + \frac{1}{\varrho}\,\frac{\partial A_\varphi}{\partial \varphi} + \frac{\partial A_z}{\partial z}$$

$$\mathrm{rot}\,\vec{A} = \vec{e}_\varrho\left(\frac{1}{\varrho}\,\frac{\partial A_z}{\partial \varphi} - \frac{\partial A_\varphi}{\partial z}\right) + \vec{e}_\varphi\left(\frac{\partial A_\varrho}{\partial z} - \frac{\partial A_z}{\partial \varrho}\right) + \frac{\vec{e}_z}{\varrho}\left(\frac{\partial}{\partial \varrho}(\varrho A_\varphi) - \frac{\partial A_\varrho}{\partial \varphi}\right)$$

$$\Delta V = \frac{\partial^2 V}{\partial \varrho^2} + \frac{1}{\varrho}\,\frac{\partial V}{\partial \varrho} + \frac{1}{\varrho^2}\,\frac{\partial^2 V}{\partial \varphi^2} + \frac{\partial^2 V}{\partial z^2}$$

Spherical coordinates:

$$u_1 = r\,; \quad u_2 = \vartheta\,; \quad u_3 = \varphi\,; \quad h_1 = 1\,; \quad h_2 = r\,; \quad h_3 = r\sin\vartheta$$

$$\vec{r} = \vec{e}_r(\vartheta,\varphi)r\,; \quad d\vec{r} = \vec{e}_r\,dr + \vec{e}_\vartheta\,r\,d\vartheta + \vec{e}_\varphi\,r\sin\vartheta\,d\varphi$$

$$dv = r^2\sin\vartheta\,dr\,d\vartheta\,d\varphi$$

$$x = r\sin\vartheta\cos\varphi\,; \quad y = r\sin\vartheta\sin\varphi\,; \quad z = r\cos\vartheta$$

$$\vec{A}(\vec{r}) \;=\; \vec{e}_r(\vartheta,\varphi)A_r + \vec{e}_\vartheta(\vartheta,\varphi)A_\vartheta + \vec{e}_\varphi(\varphi)\,A_\varphi\,; \qquad V = V(r,\vartheta,\varphi)$$

$$\operatorname{grad} V \;=\; \vec{e}_r\,\frac{\partial V}{\partial r} + \frac{\vec{e}_\vartheta}{r}\,\frac{\partial V}{\partial\vartheta} + \frac{\vec{e}_\varphi}{r\sin\vartheta}\,\frac{\partial V}{\partial\varphi}$$

$$\operatorname{div}\vec{A} \;=\; \frac{1}{r^2}\,\frac{\partial}{\partial r}(r^2 A_r) + \frac{1}{r\sin\vartheta}\,\frac{\partial}{\partial\vartheta}(A_\vartheta\sin\vartheta) + \frac{1}{r\sin\vartheta}\,\frac{\partial A_\varphi}{\partial\varphi}$$

$$\operatorname{rot}\vec{A} \;=\; \frac{\vec{e}_r}{r\sin\vartheta}\left(\frac{\partial}{\partial\vartheta}(A_\varphi\sin\vartheta) - \frac{\partial A_\vartheta}{\partial\varphi}\right) + \frac{\vec{e}_\vartheta}{r}\left(\frac{1}{\sin\vartheta}\frac{\partial A_r}{\partial\varphi} - \frac{\partial}{\partial r}(rA_\varphi)\right) +$$

$$+ \frac{\vec{e}_\varphi}{r}\left(\frac{\partial}{\partial r}(rA_\vartheta) - \frac{\partial A_r}{\partial\vartheta}\right)$$

$$\Delta V \;=\; \frac{\partial^2 V}{\partial r^2} + \frac{2}{r}\,\frac{\partial V}{\partial r} + \frac{1}{r^2\sin\vartheta}\,\frac{\partial}{\partial\vartheta}\left(\sin\vartheta\,\frac{\partial V}{\partial\vartheta}\right) + \frac{1}{r^2\sin^2\vartheta}\,\frac{\partial^2 V}{\partial\varphi^2}$$

### Relations for the Gradient-Operator

$$\operatorname{grad}(UV) \;=\; U\operatorname{grad}V + V\operatorname{grad}U$$
$$\operatorname{grad}(\vec{A}\vec{B}) \;=\; (\vec{A}\operatorname{grad})\vec{B} + (\vec{B}\operatorname{grad})\vec{A} + \vec{A}\times\operatorname{rot}\vec{B} + \vec{B}\times\operatorname{rot}\vec{A}$$

$$\operatorname{grad}(\vec{e}\vec{r}) = \vec{e}; \quad \vec{e}\text{ constant vector, }\vec{r}\text{ position vector}$$

$$\operatorname{grad}U(r) \;=\; \frac{dU}{dr}\,\frac{\vec{r}}{r}$$
$$(\vec{A}\cdot\operatorname{grad})\vec{r} \;=\; \vec{A}$$

### Relations for the Divergence-Operator

$$\operatorname{div}(U\vec{A}) \;=\; U\operatorname{div}\vec{A} + \vec{A}\operatorname{grad}U$$
$$\operatorname{div}(\vec{A}\times\vec{B}) \;=\; \vec{B}\operatorname{rot}\vec{A} - \vec{A}\operatorname{rot}\vec{B}$$
$$\operatorname{div}[\vec{r}U(r)] \;=\; 3\,U(r) + r\,\frac{dU}{dr}$$
$$\operatorname{div}\operatorname{grad}U \;=\; \Delta U$$
$$\operatorname{div}\operatorname{rot}\vec{A} \;=\; 0$$
$$\operatorname{div}(\vec{e}\times\vec{r}) \;=\; 0; \quad \vec{e}\text{ constant vector}$$

Relations for the Curl-Operator

$$\text{rot}\,(U\,\vec{A}) \;=\; U\,\text{rot}\,\vec{A} + \text{grad}\,U \times \vec{A}$$

$$\text{rot}\,(\vec{A} \times \vec{B}) \;=\; (\vec{B}\,\text{grad}\,)\,\vec{A} - (\vec{A}\,\text{grad}\,)\,\vec{B} + \vec{A}\,\text{div}\,\vec{B} - \vec{B}\,\text{div}\,\vec{A}$$

$$\text{rot}\,(\vec{r}\,U(r)) \;=\; 0$$

$$\text{rot}\,(\vec{e} \times \vec{r}) \;=\; 2\vec{e}\,; \quad \vec{e}\ \text{constant vector}$$

$$\text{rot}\,\text{grad}\,U \;=\; 0$$

$$\text{rot}\,\text{rot}\,\vec{A} \;=\; \text{grad}\,\text{div}\,\vec{A} - \Delta\,\vec{A}$$

$$\text{rot}\,\text{rot}\,\text{rot}\,(\vec{e}\,U) \;=\; \vec{e} \times \text{grad}\,(\Delta U)$$

$$\text{rot}\,\text{rot}\,\text{rot}\,(\vec{r}\,U) \;=\; \vec{r} \times \text{grad}\,(\Delta U)$$

Theorems

Stokes theorem

$$\int_a \text{rot}\,\vec{A}\,d\vec{a} \;=\; \oint_C \vec{A}\,d\vec{s}$$

$$\int_a d\vec{a} \times \text{grad}\,U \;=\; \oint_C U\,d\vec{s}$$

Gauss theorem

$$\int_v \text{div}\,\vec{A}\,dv \;=\; \oint_a \vec{A}\,d\vec{a}$$

$$\int_v \text{grad}\,U\,dv \;=\; \oint_a U\,d\vec{a}$$

$$\int_v \text{rot}\,\vec{A}\,dv \;=\; \oint_a d\vec{a} \times \vec{A}$$

Green's first identity

$$\int_v (U\,\Delta V + \text{grad}\,U\,\text{grad}\,V)\,dv \;=\; \oint_a U\,\text{grad}\,V\,d\vec{a}$$

$$\int_v (\text{grad}\,U)^2\,dv \;=\; \oint_a U\,\frac{\partial U}{\partial n}\,da \quad \text{for} \quad V = U \quad \text{and} \quad \Delta U = 0$$

Green's second identity

$$\int_v (U \,\Delta V - V \,\Delta U)\, dv \;=\; \oint_a \left( U\, \frac{\partial V}{dn} - V\, \frac{\partial U}{dn} \right) da$$

$$\int_a \Delta U\, dv \;=\; \oint_a \operatorname{grad} U\, d\vec{a}$$

Stratton's theorems

$$\int_v \left( \operatorname{rot}\vec{A}\, \operatorname{rot}\vec{B} - \vec{A}\, \operatorname{rot}\operatorname{rot}\vec{B} \right) dv \;=\; \oint_a \left( \vec{A} \times \operatorname{rot}\vec{B} \right) d\vec{a}$$

$$\int_v \left( \vec{A}\, \operatorname{rot}\operatorname{rot}\vec{B} - \vec{B}\, \operatorname{rot}\operatorname{rot}\vec{A} \right) dv \;=\; \oint_a \left( \vec{B} \times \operatorname{rot}\vec{A} - \vec{A} \times \operatorname{rot}\vec{B} \right) d\vec{a}$$

# Bibliography

[1] Spiegel, Murray R.: "Theory and Problems of Vector Analysis", Schaum's Outline Series, Mc Graw Hill Book Company, 1974

[2] Edminister, Joseph A.: "Electromagnetics", Schaum's Outline Series, Mc Graw Hill Book Company, 1984

[3] Nascar, Syed A.: "2000 Solved Problems in Electromagnetics", Schaum's Solved Problems Series, Mc Graw Hill Book Company, 1992

[4] Jackson, John D.: "Classical Electrodynamics", John Wiley & Sons, 1998

[5] Jones, Douglas S.: "The Theory of Electromagnetism", Pergamon Press, 1964

[6] Smythe, William R.: "Static and dynamic electricity", Mc Graw Hill Book Company, 1968

[7] Stratton, Julius A.: "Electromagnetic Theory", John Wiley & Sons, 2007

[8] Sommerfeld, Arnold: "Electrodynamics", Academic Press, 1966

[9] Lehner, Günther: "Electromagnetic Field Theory for Engineers and Physicists", Springer Verlag, Berlin Heidelberg New York, 2010

[10] Morse, Philip, M., Feshbach, Herman: "Methods of Theoretical Physics, Part I, Part II", Mc Graw-Hill Book Company, Boston, 1953

[11] Abramowitz, Milton, Stegun, Irene A.: "Handbook of Mathematical Functions", Dover Publications, Inc. New York, 1972

[12] Moon, P., Spencer: "Field Theory Handbook", Springer Verlag, Berlin Heidelberg New York, 1971

[13] Mrozynski, Gerd: "Elektromagnetische Feldtheorie: Eine Aufgabensammlung", Teubner Verlag, Wiesbaden, 2003

# Index